Distributed Cooperative Intelligence in Microgrid Control

Jingang Lai • Xiaoqing Lu

Distributed Cooperative Intelligence in Microgrid Control

A Communication Constraint Perspective

Springer

Jingang Lai
School of Artificial Intelligence
and Automation
Huazhong University of Science
and Technology
Wuhan, Hubei, China

Xiaoqing Lu
School of Electrical Engineering and
Automation
Wuhan University
Wuhan, Hubei, China

ISBN 978-981-95-6883-3 ISBN 978-981-95-6884-0 (eBook)
https://doi.org/10.1007/978-981-95-6884-0

This Springer imprint is published by the registered company Springer Nature Singapore Pte Ltd.
The registered company address is: 152 Beach Road, #21-01/04 Gateway East, Singapore 189721, Singapore

Preface

As typical smart cyber-physical systems, renewable energy microgrids are able to achieve the integration of heterogeneous and geographically dispersed intelligent nodes (e.g., electronically interfaced distributed energy resources (DER) or distributed generator (DG) including distributed photovoltaic (PV), wind turbines (WP), energy storages (ESs), electric vehicles (EVs), and so on), which can sense/actuate, control, communicate, and interact autonomously to efficiently deliver power and electricity to consumers. With the continuous empowerment of advanced information and communication technologies (ICTs), the cyber network and energy physical network in microgrids will achieve deep integration, and all DERs will become intelligent agents with strong network information capabilities and autonomous capabilities accordingly, like an "energy agent" that can obtain environmental information in real time and perform specified tasks. In turn, the behavior of these "energy agents" depends on the properties of physical devices themselves. This way, the microgrid can be regarded as a multi-agent system where each physical DER will be born with a digital twin agent (i.e., DER agent). Accordingly, ICT-enabled DER agents will own typical characteristics of autonomy, cooperation and interaction of multi-agents, by which the stability, security, and intelligence of microgrids will be further enhanced through distributed intelligent cooperation between DER agents and their neighbors.

Furthermore, considering the inherently geographically dispersed and heterogeneous characteristics of cyber-physical microgrids, with the increasing of DERs access leading to the expansion of system scale, the cyber-enhanced distributed intelligent cooperative control structure gradually replaces the conventional centralized control structure to achieve and maintain the whole system efficient, safely and stable operation. Although the extensive use of ICTs improves the flexibility and reliability of system operation, inherent communication constraints and disturbances in cyber network will inevitably affect the control effectiveness and cooperative performance of participating DER agents. Thus, this book discusses the communication constraints, such as dynamical network topology, communication delay, uncertain communication link, stochastic noise disturbance, and limited communication bandwidth, on the operation of microgrids, and proposes corresponding distributed intelligent cooperative approaches for solving the aforementioned communication constraints to achieve voltage/frequency restoration and active/reactive power sharing of microgrids.

In Chap. 1, an introduction on cyber-physical microgrids and distributed cooperative intelligence from a communication constraint perspective is provided. Chapter 2 introduces the advanced ICT-enabled microgrid with focus on its hierarchical distributed control scheme with sparse communication networks. First, a brief introduction to the graph theory and primary local control is given. Then, a systematic derivation of the relevant equations for the distributed intelligent cooperative protocol is proposed in case of various communication constraints. Meanwhile, the control objectives of microgrids are also introduced.

In Chap. 3, a novel distributed intelligent cooperative protocol with finite time regulator is proposed to coordinate distributed energy resources in islanded microgrids. The finite time frequency control strategy involving a consensus-based distributed active power controller is designed, such that the accurate active power sharing can be achieved while frequencies of all DERs can be forced to converge to the reference value within a finite time. Then, a weighted average voltage observer based voltage control strategy containing a reactive power sharing constraint is designed, which guarantees the weighted average value of all DERs' output voltages can converge to the desired reference value. The performance analysis shows the proposed distributed intelligent cooperative protocol with finite time regulator can accurately address the issue of voltage regulation, frequency restoration, and active and reactive power sharing. Furthermore, the voltage and frequency control inputs are designed with saturation constants, which can not only suppress the transient overshoot but also guarantee the steady state stability.

In Chap. 4, a robust distributed intelligent cooperative control protocol for AC microgrids under a dynamical switching communication network with nonuniform time varying delays is developed. This chapter illustrates pinning-based frequency/voltage controllers containing a distributed voltage observer. In addition, consensus-based active/reactive controllers are applied into the proposed distributed intelligent cooperative protocols to generate values of nominal set points employed in the primary control stage for all DER agents. By employing these approaches, the frequency and weighted average output voltage of all DERs can be pinned to given values and power sharing of active and reactive power is also maintained. Finally, by considering the nonuniform time varying delay in the communication network between distributed DER agents, this chapter proves the sufficient conditions of network connectivity and delay upper bound to ensure the reliability and stability of the whole closed loop system.

In Chap. 5, a discrete time distributed intelligent cooperative protocols for AC microgrids through a sparse communication network with uncertain communication links is proposed. To achieve voltage regulation, frequency restoration as well as accurate power sharing for all DER agents, novel discrete time distributed intelligent cooperative protocols are developed for AC microgrids based on the iterative learning mechanic. The discussed controllers appropriately respond to uncertain communication links through an iterative operation manner, with only the requirement of accessing limited information of system parameters. Meanwhile, the control inputs of proposed distributed intelligent cooperative protocols are simply updated at the end of each iteration operation, such that all DER agents only need to communicate with their neighbors intermittently. Therefore, communication costs are greatly reduced, and some sufficient conditions on the system stability and robustness to the uncertainties are also derived by using tools of direct Lyapunov method, algebraic graph theory, and linear matrix inequality theory. Moreover, the desired control objectives can also be guaranteed even if all DER agents are subject to internal uncertainties and external noises, including initial voltage and/or frequency resetting errors and measurement disturbances, which then improves the system reliability and robustness.

In Chap. 6, a novel resilient distributed intelligent cooperative protocol is proposed for AC microgrids subject to additive noise and communication delay disturbances. The proposed resilient multi-agent control scheme is composed of three distributed consensus protocols, which is able to restore the output voltages and frequencies of DER agents to their reference values and achieve the optimal active power sharing property in almost sure convergence by employing a low bandwidth communication network with noise and communication delay disturbances. By means of stochastic analysis tools and algebraic graph theory, distributed consensus control protocols are designed to be employed for the secondary control layer of microgrids. On this basis, the stability criteria of the closed loop microgrid control system under noise and communication delay disturbances are deduced. As a result, the proposed novel resilient consensus protocols can well restore the derivations of voltage and frequency produced at the primary control level, meanwhile, also can well achieve the optimal power sharing even though there exist communication disturbances. Several simulation scenarios on an islanded microgrid are provided to verify the control performance.

Wuhan, Hubei, China

Jingang Lai
Xiaoqing Lu

Declaration

Competing Interests The authors have no competing interests to declare that are relevant to the content of this manuscript.

Contents

Chapter 1
Introduction

Growing concerns over environmental impacts and the global energy crisis have significantly accelerated research and development in the field of clean and renewable energy technologies. Among these, distributed photovoltaic (PV) systems, wind turbines (WP), energy storage systems (ESs), and electric vehicles (EVs) have gained considerable attention as promising solutions [1–4]. This shift in the power generation structure not only supports the transition toward sustainable development but also introduces substantial challenges for the planning and reliable operation of future power systems [5–8]. However, the integration of such distributed energy resources (DERs) is complicated by their inherent stochastic, fluctuating, and intermittent nature. Unlike conventional power generation units, most DERs exhibit low inertia and fast dynamic response. While this enables quick adaptation to changing conditions, it also limits their ability to support power quality and maintain system balance through their own regulation ability [9–11]. As a result, the increasing penetration of renewable based DERs poses serious challenges to power system safety, operational stability, and overall power quality. To handle these issues, the concept of microgrids has been introduced, which enable more effective integration and coordination of diverse DERs, thereby enhancing system resilience and power quality while alleviating stress on the central grid [12–15].

1.1 Smart Microgrid Systems

Smart microgrids are a typical cyber-physical systems that integrate a wide variety of heterogeneous intelligent nodes. These nodes include electronically interfaced distributed energy resources, including distributed generators, flexible loads, battery energy storage devices as well as monitoring and protection equipment. These devices together form an interconnected network capable of efficiently delivering stable and reliable power to end consumers [16–19]. The deployment of advanced

J. Lai, X. Lu, *Distributed Cooperative Intelligence in Microgrid Control*,
https://doi.org/10.1007/978-981-95-6884-0_1

ICTs play a essential role in enabling distributed cooperative intelligence among these distributed energy resources, which promotes distributed cooperative intelligence of all DERs, who are born with digital twin agents (i.e., DER agent) [20–23]. This way, individual DERs gain the ability to sense, actuate, control, and communicate autonomously with neighboring DER. Such intelligent interaction allows the system to dynamically optimize its operation in a distributed manner [24–27]. Therefore, ICT-enabled microgrids are able to enhance the system performance in terms of reliability, resilience, flexibility, and energy efficiency. Accordingly, smart microgrids are regarded as an effective solution for the reliable integration of large number of DER agents including renewable energy resources such as wind and solar generation and is thus has various applications such as remote villages, industrial parks, business and residential buildings and military bases, etc [28–30].

Conventionally, microgrids can be divided into AC microgrids and DC microgrids according to the corresponding electrical infrastructure. The general structure of cyber-physical microgrids consists the cyber space of communication networks and software based control, and the physics microgrid system. In normal, microgrids are connected to main grid through the conversion device at the point of common coupling (PCC) [31–33]. And the critical circuit breaker is utilized to autonomously transfer the operation of microgrids from grid connected to islanded model induced by disturbances or faults [34–37]. In this scenario, the stability of voltage and frequency within the microgrid system, as well as power balance, all hinge on the autonomous and intelligent cooperative control of DER agents [38–41]. Consequently, it is imperative to establish reliable control approaches, which can ensure efficient and robust operation of such heterogeneous and large scale cyber-physical systems. In this situation, cyber-enhanced distributed intelligent cooperative theory is anticipated to serve as a pivotal analytical tool in the design of cyber-physical microgrids [42–47].

1.2 Distributed Cooperative Intelligence of Microgrid Control Systems

To efficiently restore voltage and frequency outputs, achieve accurate active and reactive power sharing, and ensure global stability in microgrid operations, a hierarchical distributed cooperative control framework has been widely adopted. As illustrated in Fig. 1.1, this framework generally comprises multiple control layers, each operating at different timescales and with distinct functionalities. At the core, the primary control layer employs the droop control technique implemented locally on each distributed DER controller. Building upon this, the secondary control layer is utilized to achieve cyber-physical coordination, i.e., each ICT-enabled physical DER is associated with a digital twin agent, and distributed intelligence is realized among these agents via sparse communication networks [27, 48–50]. Specifically, the primary control layer, operating at the fastest timescale, is directly implemented

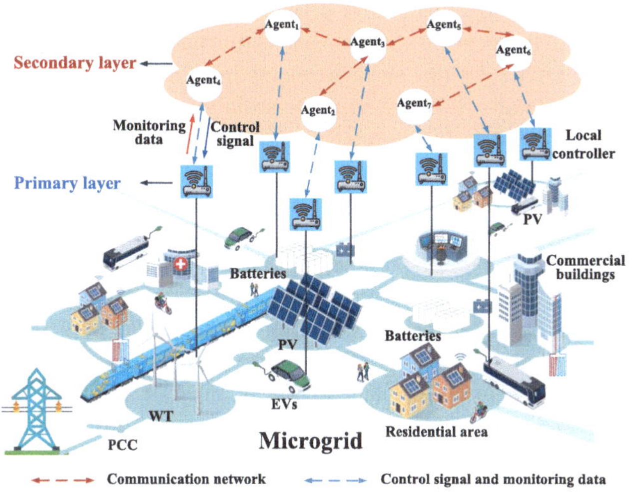

Fig. 1.1 Hierarchical distributed cooperative intelligence of cyber-physical microgrid control systems

in the inverters of DERs. It ensures the autonomy of individual DER agents to guarantee the fundamental stability across the whole microgrid system and ensure the automatic load sharing between DER agents among multiple DERs [33, 51, 52]. Despite its importance, conventional droop control technique suffers from an inherent limitation: it often leads to frequency and voltage deviations from their nominal values as load demand fluctuates. These deviations will compromise power quality and system performance. To overcome this drawback, the secondary control layer, acting at a slower timescale, is employed to eliminate the deviations introduced by the primary control layer, thereby restoring voltage and frequency to their reference values. Additionally, this layer also ensures accurate active and reactive power sharing among different DERs, enhancing the overall coordination and power quality of the microgrid [53–56]. Operating at the slowest timescale, the tertiary control layer serves as the highest decision making level. It is responsible for system level power management and optimizing energy exchange between the microgrid and the main utility grid. By considering economic criteria, grid conditions, and market signals, this layer facilitates economically optimal operation and enhances the microgrids ability to participate in broader grid-supporting services [57–60].

Recently, the control architecture of microgrid systems has undergone a significant transition, shifting away from the conventional centralized structure toward a more autonomous, flexible, and distributed framework [61, 62]. This evolution is largely driven by the requirement to accommodate the massive growth, highly dispersed and inherently unpredictable nature of DERs within modern microgrids [63, 64]. In response to these challenges, distributed cooperative control strategies based on multi-agent consensus algorithms have gradually replaced traditional centralized control methods. Such approaches have now became the mainstream architecture for achieving distributed cooperative intelligence in microgrids to achieve voltage and frequency restoration and active/reactive power sharing [65, 66]. The growing preference for this structure stems from its ability to reduce communication costs, while also offering superior scalability and operational flexibility compared to centralized architectures [67, 68]. Consequently, advanced ICTS are playing an increasingly vital role in supporting real time information exchange and distributed decision making among DER agents [69, 70]. By enabling robust and timely coordination, these technologies contribute significantly to enhancing both the resilience and the reliable operation of microgrid systems under varying operating conditions [71–73].

Normally, the cooperative intelligence control schemes are implemented through a sparse communication network. These networks play an irreplaceable role in microgrid control systems, especially as the number of connected DER agents continues to grow [74–76]. In practical applications, the utilization of advanced ICTs effectively meets the requirements of low inertia microgrid systems for coordinating fast response electronically interfaced DERs. This significantly enhances the performance of voltage regulation, frequency restoration, and load sharing within microgrid operations [77]. However, the dependency of the microgrid control system on the communication network makes it more susceptible to the unexpected communication constraints. These include communication delays [78, 79], uncertainties in communication links [80, 81], and stochastic noise disturbances [82, 83]. Such issues will severely disrupt information synchronization and compromise data accuracy, ultimately degrading the dynamic performance of the microgrid system [84, 85]. In critical cases, they may even threaten system stability [86]. Therefore, works on distributed cooperative intelligence design specifically targeting voltage regulation and frequency restoration as well as active/reactive power sharing of microgrids at the secondary control level under various communication constraints is both highly challenging and practically meaningful. Developing robust control strategies that can accommodate these aforementioned limitations remains a valuable and worthwhile direction for advancing microgrid technology and ensuring its reliable application.

1.3 Microgrid Control Systems: Topics in Communication Constraint Perspective

In practical microgrid systems, the implementation of sparse distributed communication networks among all DER agents enables accurate and timely information exchange. This capability significantly enhances system flexibility, scalability, and operational stability [87–91]. As a result, such communication architectures have become a foundational element in the design of modern distributed microgrid control systems. The adoption of advanced ICTs further supports the development of a more adaptive and resilient communication infrastructure. This contributes to improved operational efficiency and reliability in microgrids [92–95]. Nevertheless, this growing reliance on communication networks also introduces vulnerabilities. Various communication constraints will significantly compromise the stability and performance of microgrid systems. The communication constraints widely discussed in the existing relevant literature can be roughly divided into the following categories: (1) dynamical network topology [96, 97]; (2) transmission and communication delays [98, 99]; (3) uncertain communication links [100]; (4) stochastic noise disturbances [101]; (5) limited communication bandwidth (e.g., data dropouts and missing measurements) [102, 103].

In practical microgrids, such communication constraints can severely degrade cooperative control performance and, in extreme cases, lead to system instability. Therefore, it is essential to incorporate accurate models of such constraints into distributed control protocols. A well-defined modeling approach allows researchers and engineers to better understand, and quantitatively evaluate how specific communication impairments influence overall system behavior. In the sections that follow, we will provide a detailed summary and analysis of distributed cooperative intelligence protocols designed for voltage regulation, frequency restoration and active/reactive power sharing in microgrid control systems under the aforementioned communication constraints.

1.3.1 Dynamical Network Topology

The ICT-enabled distributed cooperative control scheme is adopted for the information transmission of DER agents in microgrids, thus, the efficient communication network management and the effective cross layer control as well as the better system wide situational awareness with global visibility can be enabled [104, 105]. Therefore, the expansion of microgrid scale and the plug-and-play function of DERs can be easily realized through this control framework. Although this framework promotes the flexibility and scalability of microgrids, the scale extension and the plug-and-play operation of microgrids also causes the switching characteristics for the communication topologies, which become a critical problem to be solved [106]. For instance, the plug-and-play functionality can assure the stability of the microgrid system when plugging in or unplugging out DER agents in microgrids due to the intermittent feature of renewable energy resources or unexpected disturbances [107]. In this case, the resulting communication topology of the microgrid network will switch to an inevitably different form [108]. Therefore, it is crucial for the distributed intelligent cooperative protocol to have the capability of flexibly plugging in or unplugging out DER agents in microgrid [109]. From this point of view, the cooperative control mechanism must be designed with considering the influence of the time varying communication network on the stability of the microgrid system. Recently, various literature have been studied to deal with different control problems in microgrid control system in the presence of switching topologies, including voltage regulation, frequency restoration and active/reactive power sharing [110–112].

Recently, the graph theory is widely adopted to accurately model switching communication topologies in microgrids, which would be the mainstream approach trend for designing control scheme of microgrids in the future [113–115]. Figure 1.2 shows the diagram of the dynamical network topology in microgrids, which can be summarized as four typical variations of communication topologies, including: (1) Fig. 1.2a shows a normal weighted graph; (2) Fig. 1.2b shows the variation of communication topology under the condition of link failure or removal of communication network. The breakdown of communication links among DER agents can be represented as the disconnection of the corresponding network edge; (3) Fig. 1.2c shows the variation of communication topology under the condition of unplugging out one DER in microgrids. The breakdown of communication links among the plug out DER agents and their neighbor DERs can be represented as the disconnection of the corresponding network edge; (4) Fig. 1.2d shows the variation of communication topology under the condition of plugging in a DER agent in microgrids. The recovery of communication links among the plug in DER agents and their neighbors can be represented as the reconnection of the corresponding network edge. When the communication topology of microgrids is constantly changing with the plug-and-play characteristic of DER agents, the stability of microgrid control system must be ensured under the distributed intelligent cooperative protocol [116].

In the existing work [117], the communication topology is often considered to switch arbitrarily fast. Eventually, this case will result in a large state transient at the switching time instants, which may lead to the instability of microgrids even if each switching communication network topology can produce a stable system. However, the assumption of arbitrarily fast switching communication network is inconsistent with the actual situation of plug-and-play operation. Therefore, the average dwell time method is adopted to represented the switching signals for microgrids to match the plug-and-play operation of DER agents [111]. In [106], a infinite time sequence of nonempty, bounded, and contiguous are utilized to express the slow switched characteristic of the time-varying communication network, such that the digraph is invariant during each interval and only switches at the beginning of each time interval. It should be noted that the communication network topology of microgrids is normally considered as deterministic, i.e., the corresponding adjacency matrix is a prior condition. Thus, main recent works on the distributed intelligent cooperative protocol in the secondary control level are carried out with the consideration of a deterministic fixed or time varying communication network topology.

1.3.2 Time Delay Effect

In practice, the data and information exchange with neighbor DER agents is essential for the distributed intelligent cooperative protocol in the secondary control level, where the communication between DER agents is necessary. This way,

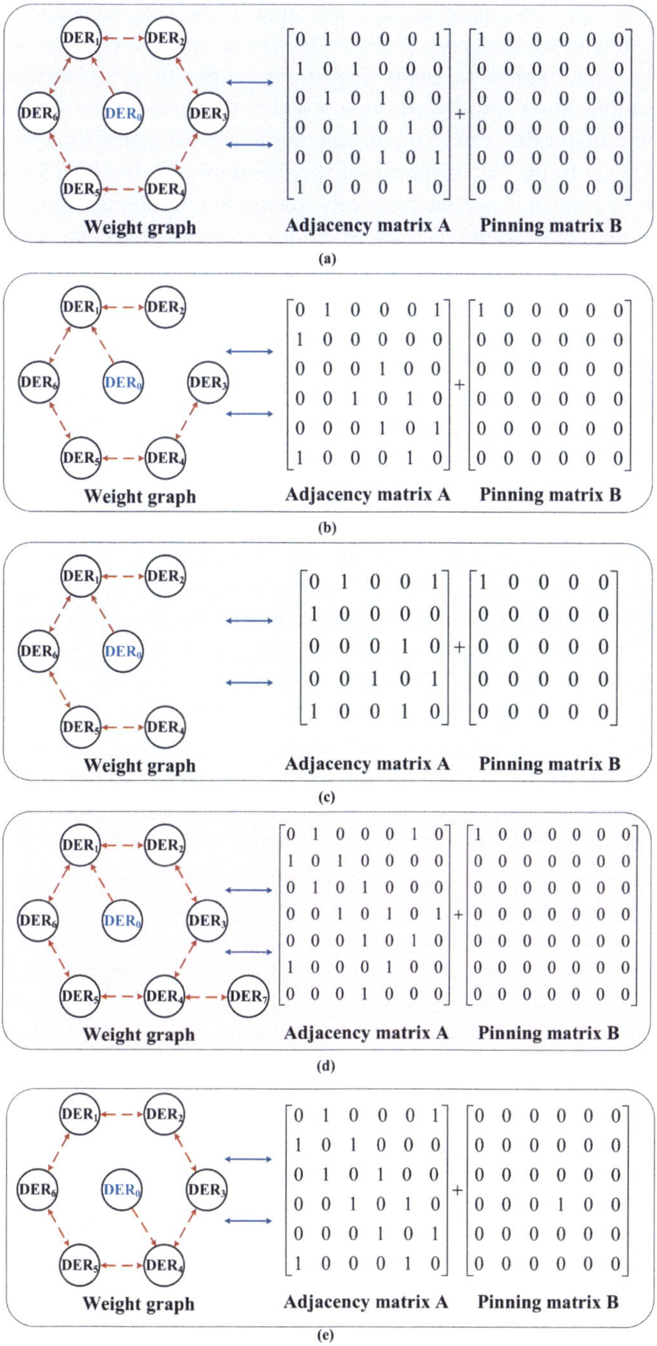

Fig. 1.2 Diagram of dynamical network topology characteristics in microgrids: (**a**) normal weight graph; (**b**) condition of link failure or removal of communication network; (**c**) condition of unplugging out a DER agent in microgrids; (**d**) condition of plugging in a DER agent in microgrids; (**e**) condition of leader connection change

inevitably introduce communication constraints in the communication networks, such as communication delays [118, 119]. There are several reasons that can result in the communication delay appearing in practical microgrids, such as limited communication speed, extra time for the measurement message reception, computation time for the control input generation, and execution time for the inputs acted [120]. Due to the fast response of the electronically interfaced DER agents to the control commands and the relatively low inertia of the microgrid system, the communication delays would have more serious impact on the control performance and stability of the microgrid control system as compared with those in bulk power systems [121]. Therefore, the communication delay with various scenarios must be considered while designing the distributed intelligent cooperative protocol in the secondary control level for practical microgrids. The communication delays in the control system of microgrids can be roughly divided into three categories [122]: (1) fixed communication delay [123], (2) time varying communication delay [124]; (3) random communication delay [125].

In practical microgrid systems, each DER agent cannot receive the real-time information from its neighbors due to the effect of communication delays. This way interrupt the efficient information sharing among DER agents and may result in oscillation, divergence, and even instability of microgrids. The fixed communication delay has a constant value which is the simplest form of communication delays. Therefore, several methods are proposed to stable the microgrid system with fixed communication delays. In [126], the maximal tolerance of a constant communication delay is given as the sufficient condition for stability of DC microgrids with constant communication delays. A novel distributed secondary control approach is proposed for the microgrid [127]. Based on this, the stability criteria of microgrid under constant delays are derived by using linear matrix inequality. In [128], a novel distributed multi-agent based finite-time secondary control strategy is proposed for the state of charge balancing and voltage regulation in microgrid with communication delays. Additionally, an important issue overlooked in mentioned literature is that the communication delay between DER agents in microgrid is usually time varying and uncertain, which cannot be handled directly by aforementioned methods.

Compared to fixed communication delays, time varying communication delays are more complex and more realistic in microgrid communication networks. To deal with time varying communication delays, robust control strategies are adopted for secondary control objectives of microgrids. For example, a stochastic distributed secondary control scheme is proposed in [129] to address the frequency restoration and accurate active power sharing problem of an AC microgrid under communication delays and noise disturbances. In order to enhance the voltage regulation and frequency restoration, a decentralized hierarchical control strategy is designed in [130] for microgrids with communication time varying delay. Furthermore, in [131], an event-triggered multistage secondary control strategy is developed for frequency deviations and power sharing errors existing in microgrids with time varying communication delays. Additionally, an important issue overlooked in mentioned literature is that all the above literature only focus on constant delays, or just consider the case that all the communication links have homogeneous time varying delays.

Due to the link breakage, message dropout and congestion of channels, the actual microgrid control system is suffering from more complicated network induced random communication delays, which is not just time varying but stochastic. Therefore, some distributed secondary control methods with delay compensation mechanism are developed to deal with the random communication delays. Recently, the generalized predictive control approach has been proposed to deal with the random communication delays due to its low requirements of Low model accuracy [132, 133]. Furthermore, the network predictive control (NPC) method has been proposed to design the delay compensation controller to enhance the control performance and robust capability against random communication delays [134]. The NPC scheme breaks through the GPC control scheme by making full use of the network characteristics, which gives it the excellent capability to compensate the random communication delay. In [134], an NPC based delay compensation controller is proposed to eliminate the adverse effect of the random communication delay to ensure the stability of microgrids.

Although the problem of delay compensation in microgrids has been extensively investigated by pertinent literature, it should be noticed that assumptions for communication delays are set as constant, time varying or random is still idealistic. In practice, the communication delay of microgrids may be more complicated subject to its own dynamics. Therefore, the distributed cooperative control designed with the consideration of communication delay determined by the system variable will be a practical and challenging problem.

1.3.3 Uncertain Communication Link

In practice, disturbances and uncertainties of communication links (e.g., network parameter uncertainties, disturbances and noises of high frequency component, data packet losses, and random communication failures) in microgrid control systems are inevitable. Therefore, it is difficult for each DER agent to get the accurate information from its neighbors due to the impact of numerical errors [135], component mismatches, transient power oscillation, high frequency disturbances and noises [136], which impacts on the stability of low-voltage microgrids are more severe than that of high-voltage ones [136, 137]. To address the defect of uncertain communication links, an iterative-learning mechanism based distributed secondary control scheme was proposed for AC microgrids through a cost-efficient sparse communication network with uncertain communication links [80]. The uncertainties of communication links are modeled that weight a_{ij} satisfy $\underline{a}_{ij} \leq a_{ij} \leq \bar{a}_{ij}$, where a_{ij} is the unknown communication weight with the upper and lower bounded $\bar{a}_{ij}$ and $\underline{a}_{ij}$, respectively.

Some of the available literature on uncertain communication links, due to link breakage or data dropout and reception of noisy or tampered data, for distributed control have been studied in [80, 138]. However, they merely are considered in the simulation case. Therefore, it is worth noting that modeling of these uncertain communication link constraints is crucial so that one can better and quantitatively understand how they affect the control performance.

1.3.4 Stochastic Noise Disturbance

In actual microgrids, communication channels are inevitably affected by noise disturbances caused by switching power supplies, silicon-controlled rectifiers, brush motors, dimmer switches, and industrial sources are directly connected to power supply system [129, 139–141]. The noise disturbances in communication channels may result in the error of information exchange among DER agents in microgrids. The adverse effects of communication noises are depicted in Fig. 1.3.

The noise disturbances can be roughly divided into two types [142–144]: (1) the additive noise; (2) multiplicative noise. First, the additive noise can be modeled as the additive Gaussian white noise, which is added to the basic deterministic part of the system state equation. It should be noticed that in practical engineering, the information from its neighbors of microgrid control systems are often be disturbed by multiplicative noises, which is considered to be an internal dynamic disturbance coupled with the system state. In practice, the multiplicative noise is more difficult to deal with than the additive noise, due to it often allows its statistical description to be unknown while depends on the real state of the system. In terms of the impact of noise on communication links, the sensitivity of microgrids to noise can reduce their dynamic performance and even lead to system instability.

Despite there are few works on this type of noises [82, 145], more efforts and research need to be focused on multiplicative noises. Finally, it is essential to filter out the unwanted noise signal from available message measurements and minimize the impacts in order to determine the best voltage/frequency restoration of microgrids. Unfortunately, it remains a challenging and open problem to design distributed filters (e.g., Kalman filters).

Fig. 1.3 Message exchange between DER$_i$ agent and DER$_j$ agent with typical communication noises: (**a**) additive noise case; (**b**) multiplicative noise case

1.4 Summary

This book discusses different distributed intelligent cooperative protocols with a cyber-enabled perspective exploited in AC microgrids under various situations of communication constraint to enhance the stability and reliability for flexible operation microgrids. The details are summarized as: Chap. 3 discusses the finite-time synchronization problem of AC microgrids under a sparse communication network. Chapter 4 discusses the synchronization problem of microgrids under a switching communication network with non-uniform time varying delays; Chap. 5 discusses the synchronization problem of AC microgrids in a distribution sparse network with uncertain communication links; Chap. 6 discusses synchronization problem of microgrids subject to additive noise and time delay disturbances.

Chapter 2
Preliminaries of Microgrid Control Systems

In recent years, the rapid advancement of communication technologies has significantly transformed information infrastructure within smart microgrid systems. Emerging technologies such as fifth generation (5G) networks and cloud computing are increasingly being adopted to enhance data transmission capabilities, thereby substantially improving both the intellectualization and interactive capacity of modern microgrid architectures. Through these advanced ICTs, DER agents are empowered to function as intelligent agents capable of autonomous sensing, actuation, control, and communication [62, 146]. These smart DER agents can continuously interact with neighboring DER agents and dynamically respond to system changes, forming a highly adaptive cyber-physical system. To properly characterize the communication relationships within this networked system, graph theory will be formally introduced in Sect. 2.1. This mathematical framework provides essential tools for describing vital communication characteristics, including connectivity, information flow paths, and network topology. Building upon this foundation, Sect. 2.2 will systematically present the control structure of microgrid systems and analyze various communication constraints that may impact microgrid system performance. These preliminaries will serve as critical elements in the subsequent design of distributed cooperative intelligence protocols, particularly for the secondary control layer of microgrids. The integration of robust theoretical frameworks with practical implementation considerations will enable the development of more resilient and efficient control strategies, ultimately supporting the reliable operation of future smart grid systems amid growing communication dependencies.

2.1 Communication Graph Theory

We discuss the basics of communication graph theory in this section. Specially, the algebraic graph theory is introduced here, which gives us a powerful and efficient

J. Lai, X. Lu, *Distributed Cooperative Intelligence in Microgrid Control*,
https://doi.org/10.1007/978-981-95-6884-0_2

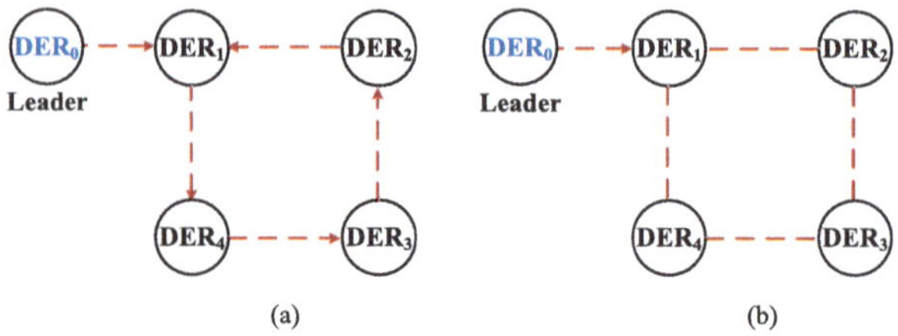

Fig. 2.1 The communication network of microgrids: (**a**) directed graph; (**b**) undirected graph

tools to analyze the ICT-enabled connection network of microgrids. The sparse peer-to-peer communication topology between multiple DER agents in microgrids can be abstracted into a graph, which can be broadly divided into the following two categories [147–149]:

(1) Directed graph as shown in Fig. 2.1a: A graph is referred to as being directed if it contains an ordered pair of vertexes. A specific direction from one vertex to another vertex is defined by the graph edge.
(2) Undirected graph as shown in Fig. 2.1b: There is no specific direction to define the edges in an undirected graph, which is a graph with an unordered pair of vertexes.

In communication networks, two-way communication is usually used in microgrids, that is to say, the information sent between DER agents and their neighbors may not be equal or same. Therefore, this assumption of the undirected graph communication is the more relaxed constraint condition than that of undirected graph communication. More details about graph theory are listed as follows.

2.1.1 Communication Graph for Microgrid Systems

From the cyber perspective, the communication network employed of microgrids can be viewed as a multi-agent system, i.e., all DER agents can be perceived to be nodes or vertices, and the sparse wireless communication network for the communication of adjacent DER agents can be regards as edges that connect nodes [150–152]. The algebraic graph theory is a mathematical object that is useful to establish the communication connection relationship among DER agents. Mathematically, a graph is a pair of sets of nodes $\mathcal{V} = \{\mathcal{V}_1, \cdots, \mathcal{V}_N\}$ and edges $\mathcal{E} = \mathcal{V} \times \mathcal{V}$, such that the graph can be defined as $G = (\mathcal{V}, \mathcal{E})$.

To be more specific about the connected relationship of DER agents in microgrids, the aforementioned graph can be further subdivided into the directed graph and the undirected graph. For a directed graph $(\mathcal{V}_i, \mathcal{V}_j) \in \mathcal{E}$ does not necessarily imply $(\mathcal{V}_j, \mathcal{V}_i) \in \mathcal{E}$, for $\mathcal{V}_i, \mathcal{V}_j \in \mathcal{V}$. On the contrary, $(\mathcal{V}_i, \mathcal{V}_j) \in \mathcal{E}$ holds if and only if $(\mathcal{V}_j, \mathcal{V}_i) \in \mathcal{E}$ holds for the undirected graph.

For a node $\mathcal{V}_i \in \mathcal{V}$ in the graph $G = (\mathcal{V}, \mathcal{E})$, the number of head ends adjacent to $\mathcal{V}_i$ is called the in-degree, which is denoted by d_i^{in}. Accordingly, the number of tail ends adjacent to $\mathcal{V}_i$ is called the out-degree, which is denoted by d_i^{out}. In microgrids, each DER$_i$ agent with itself digital twin agent is supposed to exchange information with a subset N_i of DER agents, called neighbors. The in-neighbor and out-neighbor of the vertex DER$_i$ agent are, respectively, denoted as $N_i^{\text{in}} = \{\mathcal{V}_j | (\mathcal{V}_i, \mathcal{V}_j) \in \mathcal{E}\}$ and $N_i^{\text{out}} = \{\mathcal{V}_j | (\mathcal{V}_j, \mathcal{V}_i) \in \mathcal{E}\}$. From this perspective, a graph is said to be balanced if $d_i^{\text{in}} = d_i^{\text{out}}$ for all $\mathcal{V} \in \mathcal{V}$. For the directed graph, this condition is impossible to satisfy. On the contrary, we only define the degree of node $\mathcal{V}_i \in \mathcal{V}$, the number of edges connected with $\mathcal{V}_i$. It should be noticed that every undirected graph is balanced.

For a graph $G = (\mathcal{V}, \mathcal{E})$, if there exist a path from v_i to v_j, two nodes $\mathcal{V}_i, \mathcal{V}_j \in \mathcal{V}$ are said to be connected. From this perspective, a graph $G = (\mathcal{V}, \mathcal{E})$ is said to be strongly connected, if any nodes $\mathcal{V}_i, \mathcal{V}_j \in \mathcal{V}$ are connected to each other. Therefore, the undirected graph is weakly connected instead of strongly connected.

An important weakly connected graph is a directed tree, which has the following properties:

(1) There is a zero in-degree node which is called as the root.
(2) All nodes but the root have in-degree one.
(3) The directed path from the root to each node is existed.

For a graph $G = (\mathcal{V}, \mathcal{E})$, if there exist a path from v_i to v_j, two nodes $\mathcal{V}_i, \mathcal{V}_j \in \mathcal{V}$ are said to be connected. From this perspective, a graph $G = (\mathcal{V}, \mathcal{E})$ is said to be strongly connected, if any nodes $\mathcal{V}_i, \mathcal{V}_j \in \mathcal{V}$ are connected to each other. Therefore, the undirected graph is weakly connected instead of strongly connected.

An important weakly connected graph is a directed tree, which has the following properties [150, 153]:

(1) There is a zero in-degree node which is called as the root.
(2) All nodes but the root have in-degree one.
(3) The directed path from the root to each node is existed.

Spanning Tree

In a strongly connected directed graph (or a connected undirected graph), information from any node can be reached to all nodes over the graph [154]. On the other hand, in the microgrid implementation, only one virtual leader-DER agent is required from which cannot receive information from all other DER agents, but sent the reference information to the other follower-DER agents [155]. Therefore, the graph is realized by the existence of a spanning tree without forcing the strongly connected condition to be satisfied. This idea can be summarized as below.

Considering a graph $G = (\mathcal{V}, \mathcal{E})$, and another graph $G' = (\mathcal{V}', \mathcal{E}')$. If the condition $\mathcal{V}' \subset \mathcal{V}$ and $\mathcal{E}' \subset \mathcal{E}$ are satisfied, the graph G' can be regarded as a subgraph of G with $G' \subset G$. Then, if $G' \subset G$ and $\mathcal{V}' = \mathcal{V}$, G' is a spanning subgraph of G. A spanning tree is a spanning subgraph that is a directed tree.

For the undirected graph, the existence of a spanning tree is depended on its connectivity, which can be represented by the following Lemma:

Lemma 2.1 ([156]) *An undirected graph is connected if and only if it has a spanning tree.*

Graph Laplacian

The matrix is usually utilized for the mathematical analysis of a graph. For a graph, there are several kinds of matrix are adopted for representations of a graph. We first introduce the adjacency matrix. The adjacency matrix of a graph $G = (\mathcal{V}, \mathcal{E})$ is defined as $A = [a_{ij}]$, which can be utilized to express the low bandwidth communication network employed in this chapter, and the elements a_{ij} can be represented by Li et al. [157] and Liu et al. [158]

$$a_{ij} = \begin{cases} 1, \text{ if } (v_j, v_i) \in E \\ 0, \text{ otherwise} \end{cases} \tag{2.1}$$

The size of the adjacency matrix depends on the node's number of the corresponding graph. If the number of nodes is N, the size of the adjacency matrix becomes $A \in R^{N\times N}$. The adjacency matrix indicates that if there exists an edge from $\mathcal{V}_i$ to $\mathcal{V}_j$, i.e., $(\mathcal{V}_i, \mathcal{V}_j) \in E$, then $a_{ij} = 1$, which denotes the DER$_i$ agent can receive the information from DER$_j$ agent. And if no link from $\mathcal{V}_i$ to $\mathcal{V}_j$, i.e., $(\mathcal{V}_i, \mathcal{V}_j) \notin \mathcal{E}$,then $a_{ij} = 0$. For any undirected graph, its adjacency matrix is a symmetric matrix. If $(\mathcal{V}_i, \mathcal{V}_j) \in \mathcal{E}$, then we have $(\mathcal{V}_j, \mathcal{V}_i) \in \mathcal{E}$.

Yet another important matrix for the graph representation is the Laplacian matrix. The Laplacian matrix $L = [l_{ij}]$ of a graph $G = (\mathcal{V}, \mathcal{E})$ with N nodes is defined as

$$L = D - A = \begin{cases} -a_{ij}, \text{ if } i \neq j \\ \sum_{i=1}^{N} a_{ij}, \text{ if } i = j \end{cases} \tag{2.2}$$

where A is the adjacency matrix of G, D is the degree matrix, who is a diagonal matrix with the (i, i)th element is the in-degree (or the degree for an undirected graph) d_i^{in} of node $\mathcal{V}_i \in \mathcal{V}$

$$D = \text{diag}\left\{d_1^{\text{in}}, d_2^{\text{in}}, \cdots, d_N^{\text{in}}\right\} \tag{2.3}$$

Then, we give the following Lemma:

Lemma 2.2 ([156]) *For any undirected graph G, its Laplacian matrix is symmetric.*

The eigenvalues and eigenvectors of the graph Laplacian matrix play an important role in analysis of a graph. For the eigenvalues, we have the following lemma:

Lemma 2.3 ([156]) *Let $L \in R^{N\times N}$ be the Laplacian matrix of a graph $G = (\mathcal{V}, \mathcal{E})$. Then the eigenvalues of L lie on the circle $R_L \subset C$ defined by*

$$R_L = \{z \in C : |z - \Delta| \leq \Delta\} \tag{2.4}$$

where Δ is the maximum degree and is the maximum value among the in-degrees of the nodes in $\mathcal{V}$, namely

$$\Delta = \max_{i \in V} d_i^{\text{in}} \tag{2.5}$$

From Lemma 2.3, we can easily understand that the eigenvalues of L have nonnegative real parts. This property is essential for proving the convergence of the solution to a graph-related differential equation $\dot{x} = -Lx$. The important properties of eigenvalues of the Laplacian matrix can be summarized in the following [159–161]

Let $L \in R^{N\times N}$ is the Laplacian matrix of a graph $G = (\mathcal{V}, \mathcal{E})$.

(1) The Laplacian matrix L has at least one eigenvalue of 0, and $1_N = [1, 1, \cdots, 1]^T \in R^N$ is an eigenvector of eigenvalue 0. Then, we have $L1_N = 0$.
(2) The rank of L satisfies $\text{Rank}(L) \leq N - 1$.
(3) The eigenvalues of L except for 0 lie in the open right half plane of C.
(4) The eigenvalue 0 of L is semi-simple.
(5) The eigenvalue 0 of L is simple if and only if $\text{Rank}(L) = N - 1$.

Leader Adjacency Matrix

In microgrids, the reference values of voltage amplitude and operating frequency are generated by a virtual DER agent, which is denoted by 0 and cannot receive information from all other DER agents. In the cooperative action of microgrids, the pining node is called the virtual leader-DER, and other existing nodes are called follower-DERs of that leader. Therefore, the extended digraph $\hat{G}$ is introduced to describe the interconnection topology of a microgrid consisting of one virtual leader-DER, denoted by 0, and follower-DER agents, denoted by $1, \cdots, N$ [162].

The extended digraph $\hat{G}$ is utilized to characterize the communication between DER agents including all the follower-DER agents and virtual leader-DER. The diagonal matrix $B = \text{diag}\{a_{10}, \cdots, a_{N0}\}$ is called the leader adjacency matrix is the communication weight matrix between follower-DER agents and the virtual leader-

DER, which describes whether each DER agent can directly receive information from the leader agent. The elements of the leader adjacency matrix a_{i0} is defined as following

$$a_{i0} = \begin{cases} 1, & \text{if node } i \text{ is pinned} \\ 0, & \text{otherwise} \end{cases} \tag{2.6}$$

which illustrate that if the DER_i agent can get the information from the virtual leader-DER agent, $a_{i0} = 0$, otherwise $a_{i0} = 0$.

2.1.2 Useful Lemma

Some useful lemmas are listed here, which will be used in the following chapters.

Lemma 2.4 ([163]) *If the digraph G with Laplacian matrix L is strong connected, then:*

(1) If $q = (q_1, \cdots, q_N)^T$ is a left eigenvector of L corresponding to the eigenvalue 0, then $q_i > 0$ holds for all $i = 1, \cdots, N$.
(2) All the eigenvalues of $L + B$ have positive real parts, where $B = \mathrm{a}_{10}, \cdots, a_{N0}$, $\sum_{i=1}^{N} a_{i0} > 0$ and $a_{i0} \geq 0$ for all $i = 1, \cdots, N$.

Furthermore, if the graph G is undirected, then $L + B$ is symmetrical and all the eigenvalues of the matrix$L + B$ are positive. Here, we denote these eigenvalues $0 < \lambda_1 \leq \lambda_2 \leq \cdots \leq \lambda_N$. Then we have the following lemma

Lemma 2.5 ([164]) *Each $\lambda > 0$ satisfies $\tau < 1/\lambda\bar{c}$ with $\bar{c} = \sup_{t \geq 0} c(t)$. There exists a function $\pi(t, s)$ of equation $\dot{Y}(t) = -\lambda_k c(t) Y(t - \tau)$ that satisfies*

$$\|\pi(t, s)\|^2 \leq \mu(\lambda) e^{-\eta(\lambda) \int_s^t c(u) \mathrm{d}u}, \quad t > s > 0 \tag{2.7}$$

where $\mu(\lambda)$ and $-\eta(\lambda) < \lambda(1 - \lambda\bar{c}\tau)$ are positive design parameters.

In the case where graph $\hat{G}$ is a spanning tree, the following two additional conditions are introduced.

$$\lim_{x \to \infty} \int_0^t e^{-\eta_0 \int_s^t c(u) \mathrm{d}u} c^2(s) \mathrm{d}s \tag{2.8}$$

where $\eta_0 = \min_{1 \leq j \leq N} \eta(\lambda_j)$, and $\{\lambda_j\}_{j=1,\cdots,N}$ are the positive eigenvalues of matrix $L + B$.

$$\lim_{x \to \infty} c(t) \log \int_0^t c(s) \mathrm{d}s \tag{2.9}$$

Summarized the above, such a communication network is utilized to explain the cooperative mechanism of distributed intelligent cooperative protocol in the secondary control level as detailed in following sections. Depending on the aforementioned communication topology assumptions and performance criteria, specifications of the reviewed distributed cooperative control designed in the existing literature works are based on the undirected graph communication, because this assumption is the most relaxed constraint condition, so the design of control algorithm under this condition is relatively simple. However, in practice, the two-way communication is usually used in microgrids, that is to say, the information sent among neighbor DER agents may not be equal or same. Therefore, the related research under this assumption is more in line with the actual microgrid operation, although the design of distributed control algorithm is more difficult under this condition. Therefore, the relevant research should be increased considering the bidirectional communication situation.

2.2 Hierarchical Distributed Control Structure

In this section, the primary and secondary control layers are introduced to clarify the hierarchical distributed control framework of microgrids. Aiming at the core problem of this book, models of distributed intelligence cooperative protocols with communication constraints is investigated here to quantitatively understand how they affect the control performance of microgrids. In the following, the distributed control protocols for microgrids with the consideration of communication constraints are expressed in detailed.

2.2.1 Primary Control

The droop control technique resembles the frequency regulation posed by the governor of a synchronous generator, which is usually utilized as the primary control layer. The droop control can satisfy the following requirements of microgrids [165–167]:

(1) Stabilize the voltage and frequency.
(2) Realize an automatic and decentralized load sharing.
(3) Offer the plug-and-play capability.
(4) Mitigate the circulating current.

Normally, the droop-based local controllers are employed in AC or DC microgrids can be summarized as the the following three type according to their characteristics [89, 168]:

(1) Active power-frequency droop.
(2) Reactive power-voltage droop.
(3) DC power-voltage droop.

The droop characteristics of the ith DER agent in microgrids can be mathematically generalized as

$$y_i = y_i^{\mathrm{nom}} - \eta_i \left(x_i - x_i^{\mathrm{nom}}\right) \tag{2.10}$$

where y_i and x_i are two system variables which are correlated; y_i^{nom} and x_i^{nom} are the nominal set points of the two variables, respectively; $\eta_i > 0$ is the corresponding droop coefficient of the ith DER agent.

In different local control entities (e.g. AC microgrids or DC microgrids with heterogeneous DER agents), the variables x_i and y_i as well as the parameter η_i have the different meaning, which can be summarized as Table 2.1

The conventional droop method can be implemented without communication links between DER agents in microgrids, and therefore, it cannot achieve global control objectives due to its dispersion and locality. There are some drawbacks of droop control technique as listed below [169–171]:

(1) It is impossible to satisfy multi control objectives in a local DER agent due to the one parameter of droop characteristic.
(2) The voltage is not a global quantity in microgrid, thus there exist an inherent contradiction between the voltage regulation and reactive power sharing due to the nonuniform inductance between DER agents. Thus, the reactive power control will inevitably affect the voltage regulation in microgrids.
(3) The droop control is essentially a kind of differential control, which will inevitably result the derivation of voltage and frequency from their desired values due to the variation of loads in microgrids.

Therefore, the secondary control layer is utilized to compensate the derivation of voltage and frequency caused by the primary control layer, and achieve the active/reactive power sharing. The droop characteristics for cyber-physical AC microgrids is depicted in Fig. 2.2.

Table 2.1 The typical droop characteristics of microgrids

Control Entity	x	y	η
Active power-frequency droop in AC microgrids	P_i	ω_i	m_i^P
Reactive power-voltage droop in AC microgrids	Q_i	v_i^{od}	n_i^Q
Power-voltage droop in DC microgrids	i_i	v_i	R_i^{v}

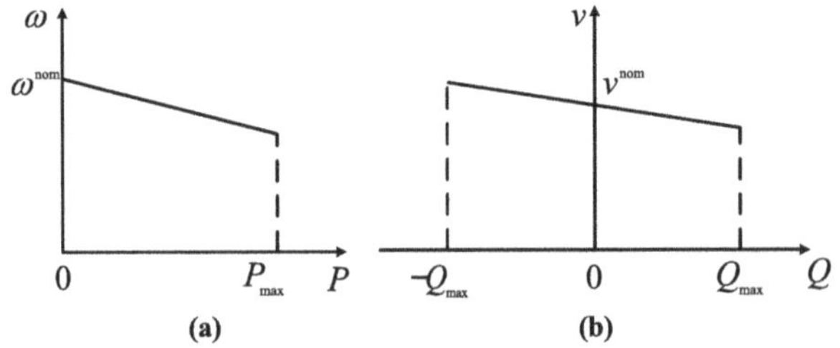

Fig. 2.2 Droop characteristics for cyber-physical AC microgrids: (**a**) voltage-reactive power droop characteristic; (**b**) frequency-active power boost characteristic

2.2.2 Distributed Cooperative Intelligence

The distributed cooperative intelligent at the secondary control layer is utilized to compensate the derivation of voltage and frequency, and achieve the active/reactive power sharing caused by the primary control layer. Specifically, the distributed intelligent cooperative protocol in the secondary control layer generates the nominal set point y_i^{nom} to synchronize the terminal output y_i to their desired value y^{DES}. In the islanded mode, the desired value y^{DES} is obtained from a virtual leader-DER. And the distributed intelligent cooperative protocol is implemented by utilizing the information from itself and exchanged with its neighbors' DER$_j$ agent ($j \in \mathcal{N}_i$) over a sparse communication network, as shown in Fig. 2.3. The empowerment of ICTs gives DER agents in microgrids to sense/actuate, control, interact autonomously and communicate with their neighbors. This way, each DER agent in the microgrid can be regarded as an agent, which can together with the other DER agents forms a multi-agent system.

The control objective of the secondary layer can be represented as: design a distributed controller for each DER$_i$ such that the terminal output y_i can be synchronized to the desired value generated by the leader-DER agent y^{DES}, i.e. $y_i \rightarrow y^{\text{DES}}$. This is known as the synchronization or cooperative tracker problem in multi-agent systemsÜnder the perfect communication network, the distributed intelligent cooperative protocol for the single-integrator dynamic is designed as following.

$$u_i(t) = \sum_{j \in \mathcal{N}_i} a_{ij}[y_j(t) - y_i(t)] + a_{i0}[y^{\text{DES}} - y_i(t)] \tag{2.11}$$

where a_{ij} is the graph edge weight and a_{i0} is the pinning gain.

With the designed distributed intelligent cooperative protocol (2.11), the single-integrator dynamic can be rewritten as

$$\begin{aligned} \dot{y}_i(t) &= \sum_{j \in \mathcal{N}_i} a_{ij}[y_j(t) - y_i(t)] + a_{i0}[y^{\text{DES}} - y_i(t)] \\ &= -y_i \sum_{j \in \mathcal{N}_i} a_{ij} + \sum_{j \in \mathcal{N}_i} a_{ij} y_j(t) + a_{i0}[y^{\text{DES}} - y_i(t)] \\ &= (d_i^{\text{in}} + a_{ij}) y_i(t) + \sum_{j \in \mathcal{N}_i} a_{ij} y_j(t) + a_{i0} y^{\text{DES}} \end{aligned} \tag{2.12}$$

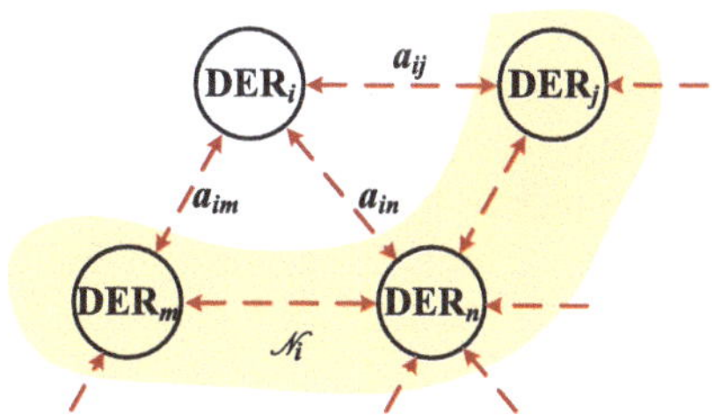

Fig. 2.3 Communication network showing the neighborhood of DER$_i$ agent, denoted by $\mathcal{N}_i$. The weight scales the message transmitted from DER$_i$ agent to DER$_j$ agent over the edge linking them

Denote the globe state vector $y(t) = [y_1(t), \cdots, y_N(t)]^T$ and the desired value vector $\tilde{y}^{\text{DES}} = 1_N y^{\text{DES}} = [y^{\text{DES}}, \cdots, y^{\text{DES}}]^T$, thus, the whole dynamic system can be expressed by

$$\begin{aligned}\dot{y}(t) &= -(D+B)y(t) + Ay(t) \\ &= -(D-A+B)y(t) + B\tilde{y}^{\text{DES}} \\ &= -(L+B)y(t) + B\tilde{y}^{\text{DES}}\end{aligned} \tag{2.13}$$

where D is the diagonal matrix of in-degrees; B is the leader adjacency matrix; A is the weight adjacency matrix.

It can be observed from (2.13) that the cooperative tracker problem is all depends on the properties of the Laplacian matrix L and leader adjacency matrix B. For simplicity of analysis, the synchronization tracking error is defined as

$$\delta(t) = \tilde{y}^{\text{DES}} - y(t) \tag{2.14}$$

where $\delta(t) = [\delta_i(t), \cdots, \delta_i(t)]^T$ is the synchronization tracking error vector.

In microgrid, the desired values of voltage and frequency are all constant. Therefore, we have

$$\dot{y}^{\text{DES}} = 0 \tag{2.15}$$

Then, the dynamic of the synchronization tracking error can be obtained from (2.13), (2.14), and (2.15) that

$$\dot{\delta}(t) = \dot{\tilde{y}}^{\text{DES}} - \dot{y}(t) = (L+B)y(t) - B\tilde{y}^{\text{DES}} \tag{2.16}$$

Since $L\tilde{y}^{\text{DES}} = 0$, we have

$$\dot{\delta}(t) = (L+B)y - (L+B)\tilde{y}^{\text{DES}} = -(L+B)\delta \tag{2.17}$$

Therefore, the convergence of the global synchronization error is determined by the properties of the matrix $L+B$. The cooperative tracking for the first-order system is guaranteed by the distributed distributed cooperative tracking protocol (2.11) if and only if the graph has a spanning tree. We shall see that all eigenvalues of $L+B$ are positive under certain graph with spanning tree, i.e., $L+B > 0$, such that

$$\dot{\delta}(t) \leq -\lambda_{\min}(L+B)\delta \tag{2.18}$$

where $\lambda_{\min}(L+B)$ is the minimum eigenvalue of matrix $L+B$.

Therefore, the system (2.17) is a stable system.

Summarize the above, the whole control structure of distributed cooperative intelligence for DER$_i$ agent with multi-loop control in an AC microgrid is depicted

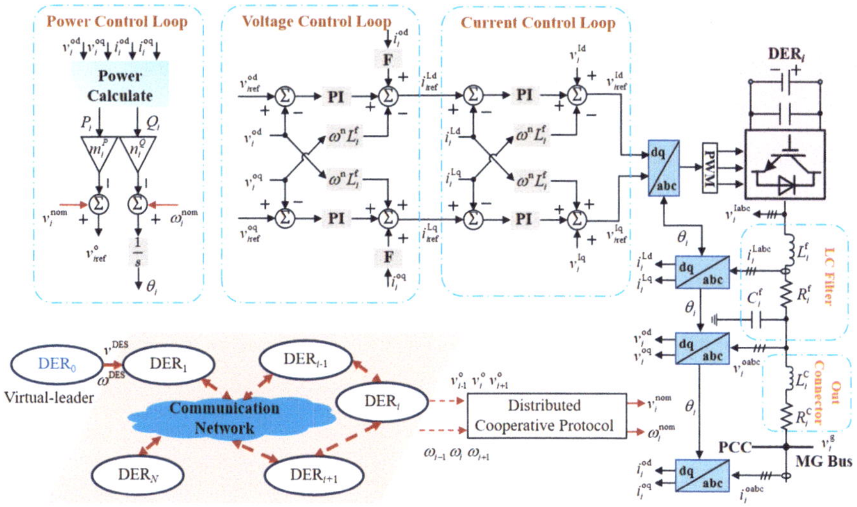

Fig. 2.4 Whole control structure of the distributed consensus control for DER$_i$ agent with multi-loop control in an AC microgrid, where the red dotted lines express the information exchange of among DER agents

in Fig. 2.4. As shown in Fig. 2.4, the hierarchical distributed control procedure is implemented under the common direct-quadrature (d-q) reference frame transformation, where d- and q-axis for the DER$_i$ agent are rotating with the common reference frequency. The basic control diagram of primary control consists of the power, voltage and current control loops. The droop control technique in the primary control layer is implemented through the power control loop with the nominal set points v_i^{nom} and ω_i^{nom}. Then, the current references $i_{i\mathrm{ref}}^{\mathrm{Ld}}$ and $i_{i\mathrm{ref}}^{\mathrm{Lq}}$ are generated by the voltage control loop with the voltage references $v_{i\mathrm{ref}}^{\mathrm{od}}$ and $v_{i\mathrm{ref}}^{\mathrm{oq}}$ provided by the power control loop. Finally, the modulated signal using to regulate the inverter outputs is generated by the current control loop with the current tracking errors. Furthermore, the required active and reactive power is calculated through a low-pass filter, where ω_i^{c} is the cutoff frequency.

To compensate the derivation of output voltage amplitude and operating frequency caused by the primary control procedure, the distributed intelligent cooperative protocol is utilized to generate the corresponding nominal set points v_i^{nom} and ω_i^{nom} through a distributed way. Specifically, a consensus-based distributed controller is designed by the exchanged information v_{i+1}^{od}, v_{i-1}^{od}, ω_{i+1}, and ω_{i-1} from the neighboring DER$_{i-1}$ agent and DER$_{i+1}$ agent to generate the nominal set points v_i^{nom} and ω_i^{nom} for the primary control procedure to further synchronize the output voltage amplitude v_i and operating frequency ω_i to their desired values v^{DES} and ω^{DES} (generated by a virtual leader DER$_0$) agent.

2.2.3 *Communication Constraint for Microgrid Systems*

However, the distributed intelligent cooperative protocol under the perfect communication network in (2.11) cannot be implemented in practice, due to the communication constraints, including the dynamical network topology [172, 173], communication delay [174, 175], uncertain communication link [176, 177], stochastic noise disturbance [178, 179] and limited communication bandwidth influence [180]. In this case, we must faced with a more complex consensus control law but a more realistic cyber-physical microgrid systems. In the following, the distributed control protocols for microgrids with the consideration of communication constraints are expressed in detailed.

Dynamical Network Topology

For the dynamical network topology, we first denote a piecewise switching signal $\sigma(t) : [0, +\infty) \rightarrow \rho_\Gamma = \{1, 2, \cdots, m\}$ defined on m digraphs $\{\hat{G}_1, \hat{G}_2, \cdots, \hat{G}_m\}$. It's clear that the communication topology is fixed if the switching signal is a constant. Figure 2.5 shows the original graph $\hat{G}_1$ in a pattern and other three patterns. If the communication link between DER_1 agent and DER_3 agent, DER_2 agent and DER_4 agent, and DER_2 agent and DER_5 agent are break due to disturbances, then, the other backup communication links between DER_3 agent and DER_4 agent, and DER_4 agent and DER_5 agent are built. It indicates that the original graph $\hat{G}_1$ is transformed into a pattern with $\hat{G}_2$. It can be observed from the Fig. 2.5 that the communication topology of microgrids can be switched among four patterns.

With the consideration of arbitrarily fast switching communication topologies, the distributed cooperative control protocols (2.11) will become

$$u_i(t) = \sum_{j \in \mathcal{N}_i} a_{ij}^{\sigma(t)}[y_j(t) - y_i(t)] + a_{i0}^{\sigma(t)}[y^{\mathrm{DES}} - y_i(t)] \tag{2.19}$$

where $a_{ij}^{\sigma(t)}$ is the change of graph edge weight from DER_i agent to DER_j agent at time $\sigma(t)$ and $a_{i0}^{\sigma(t)}$ is the change of pinning gain of DER_i agent at time $\sigma(t)$.

Time Delay Effect

The communication delay is an inherent characteristic of the communication network in microgrid systems. Due to the fast response of the electronically interfaced DER agents to the control commands and the relatively low inertia of the microgrid system, the communication delays would have more serious impact on the control performance and stability of the microgrid control system as compared with those in bulk power systems. Therefore, the communication delay with various

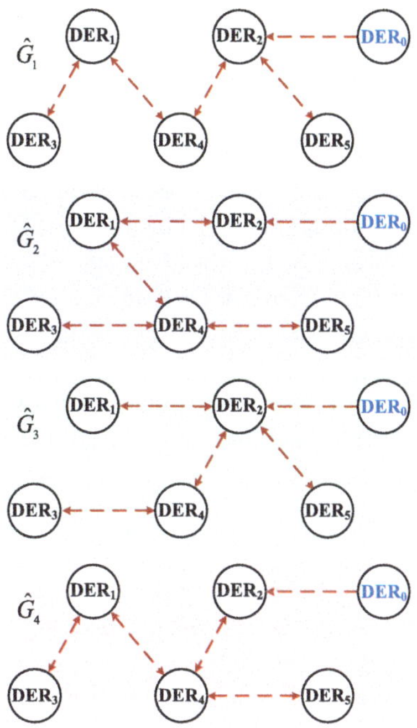

Fig. 2.5 Switching communication topologies of microgrids

scenarios must be considered while designing the distributed intelligent cooperative protocol for practical microgrid system. The communication delays in the microgrid control system of microgrids can be considered as: (1) symmetric and asymmetric communication delay, (2) aperiodic sampling delay;

Denote $\tau(t)$ is the communication delay, which expresses the time it takes for information to be transmitted from DER_j agent to DER_i agent. The distributed intelligent cooperative protocol considering asymmetric communication delays is designed as

$$u_i(t) = \sum_{j\in\mathcal{N}_i} a_{ij}[y_j(t-\tau(t)) - y_i(t)] + a_{i0}[y^{\text{DES}} - y_i(t)] \tag{2.20}$$

where $y_i(t)$ is the measured information variable from DER_i agent itself and $y_j(t-\tau(t))$ is the measured information variable with communication delay $\tau(t)$ from DER_i agent's neighbor DER_j agent. The DER_i agent cannot receive the real-time information $y_j(t)$ at time t due to the existence of communication delays. However, the aforementioned cooperative protocol (2.20) ignores the computation time and execution time of DERi agent. Hence, the DER_i agent cannot receive its own real-time information at time t either. Therefore, the distributed intelligent cooperative protocol considering symmetric communication delays is designed as

$$u_i(t) = \sum_{j\in\mathcal{N}_i} a_{ij}[y_j(t-\tau(t)) - y_i(t-\tau(t))] + a_{i0}[y^{\text{DES}} - y_i(t-\tau(t))] \tag{2.21}$$

where $y_i(t-\tau(t))$ and $y_j(t-\tau(t))$ are the measured information variables with communication delays from DER$_i$ agent itself and its neighbor DER$_j$ agent. It should ne noticed that if the communication delay $\tau(t)$ is a constant, the corresponding communication delay is fixed, otherwise, the corresponding communication delay is time varying.

The aforementioned distributed intelligent cooperative protocol will result in high communication burden due to the continuous-time communicating mechanism. To overcome this disadvantage, the sampled-data-based communicating mechanism is adopted here for DER agents' information exchange. Specifically, the DER$_j$ agent will send its measured information variable with time stamp to the DER$_i$ agent with aperiodic sampling times ς_k. And a buffer is utilized to collect the received information to ensure the orderliness of the data under communication delays. The aperiodic sampling times is defined as

$$0 = \varsigma_0 < \varsigma_1 < \cdots < \varsigma_k < \cdots, \lim_{k\to\infty} \varsigma_k = \infty \tag{2.22}$$

with the interval between adjacent sampling time ς_{k+1} and ς_k is denoted as $h_k = \varsigma_{k+1} - \varsigma_k$, which satisfies $0 \le \underline{h} \le h_k \le \bar{h}$, where $\underline{h}$ and $\bar{h}$ is the lower and upper bounded of the interval h_k, respectively.

With the sampled-data-based communicating mechanism, the time varying communication delay in the distributed intelligent cooperative protocol is

$$\tau(t) = t - t_k + \underline{h} \tag{2.23}$$

where t_k is the moment when the controllers (2.21) update its inputs.

The time varying communication delay $\tau(t)$ is related to the aperiodic sampling time and communication delay $\theta_k(t)$. Specifically, the DER$_j$ agent measures the information at the aperiodic sampling instant ς_k, and then submit it to its neighbor DER$_i$ agent. During the transmission, the DER$_i$ agent receives the after time θ_k due to the transmission delay in the communication network. In order to ensure the orderliness of the data under communication delays, a buffer is adopted here to further delay the received messages with the waiting time $\underline{h}-\theta_k$. Therefore, the time varying communication delay in the distributed intelligent cooperative protocol with the sampled-data-based communicating mechanism is $\tau(t) = t - t_k + \underline{h}$.

Uncertain Communication Link

In practice, the disturbances and uncertainties of communication links (e.g., network parameter uncertainties, disturbances and noises of high frequency component, data packet losses, and random communication failures) in the microgrid control system are inevitable. Therefore, it is difficult for each DER agent to get the accurate information from its neighbors due to the impact of numerical errors [135], component mismatches, transient power oscillation, high frequency disturbances and noises [136], which impacts on the stability of low-voltage microgrids are more severe than that of high-voltage ones [136, 137].

For the uncertain communication link, it can be assumed that each nonzero elements of the weighted adjacency matrix a_{ij} satisfies $0 < \underline{a}_{ij} \leq a_{ij} \leq \overline{a}_{ij}$, where $\underline{a}_{ij}$ and $\overline{a}_{ij}$ are the nonnative bounds. Denote $\underline{A}$ and $\overline{A}$ are the corresponding lower bound matrix and upper bound matrix, which respectively correspond to the digraph $G(\underline{A})$ and $G(\overline{A})$. It is obvious that $G(\underline{A})$ is a spanning sub-graph of $G(\overline{A})$, and the associated neighbor sets of the ith DER agent satisfies $\underline{\mathcal{N}_i} \subseteq \mathcal{N}_i \subseteq \overline{\mathcal{N}_i}$.

Due to the graph edge weight a_{ij} is unknown, the aforementioned distributed intelligent cooperative protocol (2.11) cannot be achieved through the general way

$$u_i(t) = \sum_{j \in \mathcal{N}_i} a_{ij}[y_j(t) - y_i(t)] + a_{i0}[y^{\mathrm{DES}} - y_i(t)] \tag{2.24}$$

Stochastic Noise Disturbance

In actual microgrids, the communication channels are inevitably affected by the noise disturbances caused by switching power supplies, silicon-controlled rectifiers, brush motors, dimmer switches, and industrial sources are directly connected to the supply network [129, 139, 140]. The noise disturbances in communication channels may result in the error of information exchange among DER agents in microgrids. In a noise environment, the designed distributed intelligent cooperative protocol (2.11) will be affect by additive stochastic noises, which can be expressed as

$$u_i(t) = \sum_{j \in \mathcal{N}_i} a_{ij}[y_j(t) - y_i(t) + \sigma_{ij}\varpi_{ij}] + a_{i0}[y^{\mathrm{DES}} - y_i(t) + \sigma_i \varpi_i] \tag{2.25}$$

where ϖ_{ij} is the measurement noises and $\sigma_{ij} > 0$ is the Gaussian white noise.

It should be noticed that in practical engineering, the information from its neighbors of microgrid control systems are often be disturbed by multiplicative noises, which is considered to be an internal dynamic disturbance coupled with the system state. In practice, the multiplicative noise is more difficult to deal with than the additive noise due to it often allows its statistical description to be unknown while depends on the real system state. Thus, the received information of DER_i agent from the DER_j agent can be represented by

$$\xi_{ji}(t) = y_j(t) + f_{ji}(y_{ji}(t))\varpi_{ji}(t) \tag{2.26}$$

where $y_{ji}(t) = y_j(t) - y_i(t)$ and $f_{ji}(y_{ji}(t))$ is the multiplicative noise.

Therefore, the distributed intelligent cooperative protocol with the consideration of multiplicative noise can be expressed by

$$u_i(t) = \sum_{j \in \mathcal{N}_i} a_{ij}[\xi_{ji}(t) - y_i(t)] + a_{i0}[y^{\mathrm{DES}} - y_i(t)] \tag{2.27}$$

2.3 Summary

In this chapter, we have introduced the microgrid with focus on its hierarchical control with the advanced ICT-enabled communication network. First, a brief introduction to the graph theory and primary control is given. Later, a systematic derivation of the relevant equations for the distributed intelligent cooperative protocol is proposed in case of various communication constraints. And the control objectives for primary and secondary level in microgrids are also introduced.

Chapter 3
Distributed Cooperative Intelligence for AC Microgrids with Finite Time Regulator

In this chapter, a novel distributed intelligent cooperative protocol with finite time regulator is proposed to coordinate distributed energy resources in islanded microgrids. The finite time frequency control strategy involving a consensus-based distributed active power controller is designed, such that the the accurate active power sharing can be achieved while frequencies of all DERs can be forced to converge to the reference value within a finite time. Then, a weighted average voltage observer based voltage control strategy containing an reactive power sharing constraints is designed, which guarantee the weighted average value of all DERs'output voltages can converge to the desired reference value. The performance analysis shows the proposed distributed intelligent cooperative protocol with finite time regulator can accurately address the issue of voltage regulation, frequency restoration, active and reactive power sharing. Furthermore, the voltage and frequency control inputs are designed with saturation constants, which can not only suppress the transient overshoot, but also guarantee the steady-state stability.

Most of the exist literature investigates the distributed cooperative intelligence issue of voltage regulation and frequency restoration as well as active power sharing with the asymptotically convergence speed [40, 181]. In order to increase the convergence speed, a finite time frequency controller and a finite-time approximate consensus strategy are proposed in [182] and [183], respectively. However, the voltage regulation and reactive power sharing are not considered in the above literature. To deal with this problem, the finite time voltage controller is investigate in [77], through which the voltage controller and frequency controller can be designed separately. Nevertheless, the accurate reactive sharing problem is still not be considered.

To our best knowledge, few literature investigates the accurate reactive power sharing problem in microgrids with line impedances, due to the inherent contradiction of voltage regulation and reactive power sharing [184]. In [185], an enhanced demand sharing method is designed by utilizing estimated reactive power

J. Lai, X. Lu, *Distributed Cooperative Intelligence in Microgrid Control*,
https://doi.org/10.1007/978-981-95-6884-0_3

errors obtained by injecting small active power disturbances. The proposed method is implemented by a central controller without relying on communication links between DERs. Then, a virtual impedance optimization method was also presented to reduce the reactive power errors [186]. Furthermore, an improve reactive power sharing strategy is proposed in [187] by combining the adaptive virtual impedance method and communication. And a harmonic power sharing scheme is utilized in case of power imbalance [188]. With the consideration of microgrid with line impedances, a novel reactive power sharing approach is proposed in [40], where the averaged value of all DERs' measured reactive powers must be estimated as nominal states. This way, the reactive power sharing can be achieved independently from line impedances and voltage mismatches.

This chapter investigates the voltage regulation, frequency restoration, active and reactive power sharing in islanded microgrids and proposes a finite time distributed cooperative intelligent protocol. The proposed finite time frequency controller ensures the frequency restoration and accurate active power sharing can be achieved within finite time, and the designed voltage controller guarantees the voltage regulation and reactive power sharing in a slower time scale. Compared to other control scheme based on the asymptotic convergence stability [40, 80, 189], the proposed method can Reduce the inherent coupling of voltage and frequency, which in turn permits us to design these controllers separately. An observer based voltage control strategy is proposed to regulate weighted average voltages of all DERs to converge to their reference value, thus the accurate reactive power sharing in presence of strongly nonuniform line impedances can be achieved through a compromise strategy. Therefore, the proposed active power control scheme is different from the schemes [77, 80], which did not consider the voltage regulation or reactive power sharing. And all controllers are designed with saturation constriction such that the transient overshoot can be greatly suppressed and the steady-state stability of the control system can be guaranteed.

3.1 Distributed Cooperative Intelligence with Finite Time Regulator

Consider an islanded microgrid with N DERs, each DER consists of a DC source, a DC/AC converter, an inductance-capacitance-inductance (LCL) filter and a output connector with resistance and inductance. For the ith DER, primary and secondary procedure are depicted in Fig. 2.4.

As seen, the entire primary control process adopts a three loop scheme of power, voltage, and current. The power control loop is implemented through the nominal set points v_i^{nom} and ω_i^{nom}, which are generated by the designed distributed cooperative intelligence. Then, the voltage control loop generate current reference value $i_{i\mathrm{ref}}^{\mathrm{Ld}}$ and $i_{i\mathrm{ref}}^{\mathrm{Lq}}$ through voltage reference value $v_{i\mathrm{ref}}^{\mathrm{od}}$ and $v_{i\mathrm{ref}}^{\mathrm{oq}}$. Finally, current errors are calculated to regulate the output of the inverter by the sinusoidal pulse width modulation (SPWM) mode [190, 191].

The droop control technique are commonly utilized in the primary control layer and can be expressed by

$$\begin{cases} v_i = v_i^{\text{nom}} - n_i^Q Q_i \\ \omega_i = \omega_i^{\text{nom}} - m_i^P P_i \end{cases} \tag{3.1}$$

where v_i and ω_i are output voltage magnitude and operating frequency of the ith DER; v_i^{nom} and ω_i^{nom} are nominal set points for DER$_i$'s voltage and frequency; m_i^P and n_i^Q are droop coefficients, which are selected according to output power ratios of DER or the formulas $m_i^P = \Delta\omega/P_{i,\max}$ and $n_i^Q = \Delta v/Q_{i,\max}$, with the acceptable maximum derivations of frequency and voltage $\Delta\omega$ and Δv [190]; $v_i = \sqrt{(v_i^{\text{od}})^2 + (v_i^{\text{oq}})^2}$ is the output voltage magnitude with v_i^{od} and v_i^{oq} are the direct and quadrature components of output voltage, respectively; P_i and Q_i are the measured output active and reactive power after the low-pass filters, which can be calculated by

$$\begin{cases} P_i = \dfrac{\omega_i^{\text{c}}}{\omega_i^{\text{c}} + s} P_i^{\text{m}} = \dfrac{\omega_i^{\text{c}}}{\omega_i^{\text{c}} + s} (v_i^{\text{od}} i_i^{\text{od}} + v_i^{\text{oq}} i_i^{\text{oq}}) \\ Q_i = \dfrac{\omega_i^{\text{c}}}{\omega_i^{\text{c}} + s} Q_i^{\text{m}} = \dfrac{\omega_i^{\text{c}}}{\omega_i^{\text{c}} + s} (v_i^{\text{od}} i_i^{\text{oq}} - v_i^{\text{oq}} i_i^{\text{od}}) \end{cases} \tag{3.2}$$

where ω_i^{c} is the cutoff frequency of the first-order low-pass filter installed in the ith DER's power control loop; i_i^{od} and i_i^{oq} are the direct and quadrature components of operating frequency, respectively; P_i^{m} and Q_i^{m} are the instantaneous active and reactive power components calculated from the measured output voltage and output current.

Since the output voltage magnitude is aligned on the d-axis of reference frame of the ith DER, we have $v_i^{\text{oq}} = 0$, then

$$\begin{cases} v_i^{\text{od}} = v_i^{\text{nom}} - n_i^Q Q_i \\ \omega_i = \omega_i^{\text{nom}} - m_i^P P_i \end{cases} \tag{3.3}$$

In order to compensate voltage and frequency derivations caused by the primary control layer, the distributed cooperative intelligence with finite time regulator will be designed, which only calculated through it's own and neighbors' information from the sparse communication network, and then generate nominal set points of voltage and frequency v_i^{nom} and ω_i^{nom} for the primary control process. A virtual leader DER is introduced to specify reference values for voltage v^{DES} and frequency ω^{DES}, which can only be accessed by a small part of DERs or at least one DER.

And all DERs can access them indirectly by the information exchanges through the communication network. Therefore, terminal outputs (voltage/frequency) of each DER can be finally regulated to the reference value.

The basic control objective is regulate voltages and frequencies of all DERs to their desired values. According to (3.3), the dynamic of voltage can be obtained

$$\dot{v}_i = \dot{v}_i^{\mathrm{nom}} - n_i^Q Q_i = u_i^v \tag{3.4}$$

where u_i^v is control input of voltage.

Accordingly, the dynamic of frequency is

$$\dot{\omega}_i = \dot{\omega}_i^{\mathrm{nom}} - m_i^P P_i = u_i^\omega \tag{3.5}$$

where u_i^ω is control input of frequency.

Then, the nominal set points of voltage and frequency can be calculated by

$$\begin{aligned} v_i^{\mathrm{nom}} &= \int_{t_0}^{t} u_i^v(\tau) + m_i^P u_i^P(\tau)\mathrm{d}\tau \\ \omega_i^{\mathrm{nom}} &= \int_{t_0}^{t} u_i^\omega(\tau) + n_i^Q u_i^Q(\tau)\mathrm{d}\tau \end{aligned} \tag{3.6}$$

where $u_i^P = \dot{P}_i$ and $u_i^Q = \dot{Q}_i$ are control inputs of active and reactive power, respectively.

In order to promote the decoupling of voltage and frequency to improve the system performance, all phase angles must be decreased to small values as fast as possible. In special, the frequency restoration and active power sharing are achieved within finite time, then we regulate the voltage while maintaining the reactive power sharing in a slow time scale. Therefore, the inherent system coupling can be further reduced, and controllers of voltage and frequency can be designed separately. Furthermore, to achieve the reactive power sharing can be achieved in microgrids with line impedances, a compromise strategy in which weighted average values of voltages of DERs converge to the desired reference value is adopted.

Summarize the above, basic control objectives of this chapter is to design the distributed cooperative intelligent protocol u_i^v, u_i^ω, u_i^P and u_i^Q to regulate the nominal set points v_i^{nom} and ω_i^{nom}, such that:

The weighted average values of voltages of all DER can be regulated to their desired value v^{DES} asymptotically

$$\lim_{t \to +\infty} \left| \sum_{i=1}^{N} \mu_i v_i(t) - v^{\mathrm{DES}} \right| = 0 \tag{3.7}$$

where $\mu = (\mu_1, \cdots, \mu_N)^T$ is the normalized positive weighting factor.

While the accurate reactive power sharing can be maintained

$$\lim_{t \to +\infty} \left| \sum_{i=1}^{N} \frac{Q_i}{Q_{i,\max}} - \frac{Q_j}{Q_{j,\max}} \right| = 0, \forall i \neq j \tag{3.8}$$

The frequency restoration can be achieved within finite time t_ω^*, i.e.

$$\lim_{t \to t_\omega^*} \left| \sum_{i=1}^{N} \omega_i(t) - \omega^{\text{DES}} \right| = 0, \omega_i(t) = \omega^{\text{DES}}, \forall t \geq t_\omega^* \tag{3.9}$$

while the accurate active power sharing can be maintained within finite time t_P^*

$$\lim_{t \to t_P^*} \left| \sum_{i=1}^{N} \frac{P_i}{P_{i,\max}} - \frac{P_j}{P_{j,\max}} \right| = 0, \forall i \neq j \tag{3.10}$$

where convergence time t_ω^* and t_P^* satisfies $t_\omega^*, t_P^* \geq t_0$.

3.1.1 *Finite Time Frequency Restoration and Active Power Sharing*

Across the sparse communication network, the global consensus error of frequency is defined as

$$e_i^\omega = \sum_{j \in \mathcal{N}_i^\omega} a_{ij}^v (\omega_j - \omega i) + a_{i0}^\omega (\omega^{\text{DES}} - \omega_i) \tag{3.11}$$

where $A^\omega = (a_{ij}^\omega)_{N \times N}$ is the adjacency matrix and $B^\omega = \text{diag}\{a_{10}^\omega, \cdots, a_{N0}^\omega\}$ with $a_{i0}^\omega \geq 0$ is the leader adjacency matrix in graph G^ω, which represent the communication network of frequency. It is assumed that there exist at least one a_{i0}^ω is not 0, such that DER$_i$ can receive the reference information ω^{DES}.

And the the global error of active power for the ith DER is

$$e_i^P = \sum_{j \in \mathcal{N}_i^P} a_{ij}^P \left(\frac{P_j}{P_{j,\max}} - \frac{P_i}{P_{i,\max}} \right) \tag{3.12}$$

where $A^P = (a_{ij}^P)_{N \times N}$ is the adjacency matrix in G^P, which represent the communication network of active power.

In order to achieve finite time frequency restoration and active power sharing, the sign function and the fractional power integrator are adopted here for controller design [192]. Furthermore, to avoid the transient overshoot and guarantee the steady-state stability of the control system, the saturation constraint is introduced to bound control inputs. Based on the above, the distributed cooperative intelligent protocol for finite time frequency restoration and active power sharing are designed as following

$$\begin{cases} u_i^{\omega} = \text{sat}_{\delta_\omega}\left[k\text{sig}(e_i^{\omega})^{\frac{1+\alpha}{2}}\right] + \hat{u}_i \\ \dot{\hat{u}}_i = \text{sat}_{\delta_\omega}\left[\gamma\text{sig}(e_i^{\omega})^{\alpha}\right], i = 1, 2, \cdots, N \end{cases} \tag{3.13}$$

$$u_i^P = P_{i,\max}\text{sat}_{\delta_P}\left[k\text{sig}(e_i^P)^{\alpha} + \gamma e_i^P\right], i = 1, 2, \cdots, N \tag{3.14}$$

where k and γ are positive control gain; $0 < \alpha < 1$ is the fractional power constant; δ_ω and δ_P are saturation constraints for frequency and active power control inputs, respectively.

Then, we can establish the following theorem.

Theorem 3.1 *If the graph G^P and the detail-balanced digraph G^ω contain a spanning tree, respectively. The frequencies of all DERs can be regulated to their desired value ω^{DES} within finite time t_ω^* under the designed distributed cooperative intelligent protocols* (3.13). *And the accurate active power sharing can be achieved within finite time t_P^* under the designed distributed cooperative intelligent protocols* (3.14).

Proof Denote $\bar{\omega}_i = \omega_i - \omega^{\text{DES}}$, $\bar{\omega} = (\bar{\omega}_1, \cdots, \bar{\omega}_N)^T$, and $x = (x_1, \cdots, x_N)^T$ with $x_i = e_i^{\omega}$, then we have $x = (L^\omega + B^\omega)\bar{\omega}$. According to (3.11) and (3.13), it can be obtained that

$$\dot{x} = -k(L^\omega + B^\omega)\text{sig}^{1+\alpha/2}(x) + (L^\omega + B^\omega)y \tag{3.15}$$

where $y = (\hat{u}_1, \cdots, \hat{u}_N)^T$ with $\dot{y} = -\gamma\text{sig}^{\alpha}(x)$.

Furthermore, define $p = (p_1, \cdots, p_N)^T$ with $p_i = P_i/P_{i,\max}$, and $z = (z_1, \cdots, z_N)^T$ with $z_i = e_i^P$, then we have $z = L^P p$. Substituting (3.14) into (3.12), yields

$$\dot{z} = -k\text{sig}^{\alpha}(z) - \gamma L^P z \tag{3.16}$$

Now we establish a Lyapunov function as $E = E_1 + E_2$ with

$$E_1 = \gamma\left(|\hat{x}|^{1+\alpha/2}\right)\text{diag}(\sigma^\omega)\left(|\hat{x}|^{1+\alpha/2}\right) + \frac{1+\alpha}{2}y^T\text{diag}(\sigma^\omega)(L^\omega + B^\omega)y \tag{3.17}$$

$$E_2 = \frac{k}{1+\alpha}\left(|\hat{z}|^{1+\alpha/2}\right)\text{diag}(\sigma^P)\left(|\hat{z}|^{1+\alpha/2}\right) + \frac{1}{2}\gamma z^T z \tag{3.18}$$

where $\sigma^{\omega} = (\sigma_1^{\omega}, \cdots, \sigma_N^{\omega})^T$ is a positive vector which satisfies $(\sigma^{\omega})^T L^{\omega} = (L^{\omega})^T \sigma^{\omega}$ for G^{ω}, and $\sigma^P = (\sigma_1^P, \cdots, \sigma_N^P)^T$ is the positive left eigenvector for the zero eigenvalue of L^P corresponding to G^P.

The derivative of Lyapunov function can be calculated by

$$\dot{E} \leq -c_1[E_1(x, y)]^{(1+3\alpha)/(2+2\alpha)} - c_2[E_2(z)]^{(2\alpha)/(1+\alpha)} \leq 0 \tag{3.19}$$

where c_1 and c_2 are two positive constants related to E_1 and E_2, respectively.

It can be obtained from (3.19) that $E = \dot{E} = 0$, we can yield $(x^T, y^T, z^T)^T = 0$. Therefore, we have $x = (L^{\omega}+B^{\omega})\bar{\omega} = 0$ with a positive stable matrix $L^{\omega}+B^{\omega}$, then $\bar{\omega} = 0$ and thus $\omega_1 = \cdots = \omega_N = \omega^{\text{DES}}$. Furthermore, we have $z = -L^P p = 0$ with $\text{rank}(L^P) = N - 1$, which implies that $p = 0$ and thus $p_1 = \cdots = p_N$, i.e. $P_1/P_{1,\max} = \cdots, P_N/P_{N,\max}$. Therefore, the frequency restoration and accurate active power sharing can be achieved within finite time $t^* = \max\{t_{\omega}^*, t_P^*\}$ with

$$\begin{aligned} t_{\omega}^* &= \frac{2(1+\alpha)}{c_1(1-\alpha)}[E_1(e^{\omega}(0))]^{\frac{1-\alpha}{2(1+\alpha)}} \\ t_P^* &= \frac{2(1+\alpha)}{c_2(1-\alpha)}[E_2(e^P(0))]^{\frac{1-\alpha}{(1+\alpha)}} \end{aligned} \tag{3.20}$$

where $e^{\omega} = (e_1^{\omega}, \cdots, e_N^{\omega})^T$ and $e^P = (e_1^P, \cdots, e_N^P)^T$ are global error vectors of frequency and active power, respectively.

Remark 3.1 For the state errors in (3.11) and (3.12), the required information is not global but adjacent, which can be easily obtain from its neighbors. Therefore, the requirement of information exchange can be satisfied by a low band-width sparse communication network, which is much useful for the microgrids containing a large number of DERs.

Remark 3.2 The introduced saturation constraint effectively overcomes the problem of consequent transient overshoot and the high frequent oscillation near the equilibrium [192]. Therefore, apart from (3.13) and (3.14), the saturation constraint is also utilized in voltage and reactive power controller to bound control inputs.

Remark 3.3 It can be obtained from the main proof that the control parameters can be set without strict constraints, and the control performance can be realized if $k > 0$, $\gamma > 0$ and $0 < \alpha < 1$. From (3.20), the convergence time of frequency t_{ω}^* and active power t_P^* are determined by control parameters k, γ, the fractional power parameter α and initial state errors. Different control parameters k and γ will result in different convergence speeds. Moreover, the convergence time can also be shorted by selecting a reasonable fractional power parameter α.

Remark 3.4 The original active power distribution pattern will be affected after the distributed cooperative intelligent protocol of frequency restoration implemented. Therefore, we design the distributed cooperative intelligent protocol (3.14) with finite time regulator to further ensure the active power sharing in microgrid.

3.1.2 Voltage Regulation and Reactive Power Sharing

Due to lack of static compensation, the accurate reactive power sharing must be guaranteed to prevent overloading in small-scale microgrid with low ratings DERs [193]. However, the voltage regulation and accurate active power sharing cannot be achieved simultaneously due to the inherent contradiction caused by the line impedance effect. For this reason, we design a compromise strategy to regulate the weighted average value of all DERs' voltages to the desired value while maintaining the accurate reactive power sharing.

Firstly, we can obtain the estimate voltages of all DERs through a voltage observer, and then regulate the estimated value $\hat{v}_i$ to the desired value v^{DES} asymptotically by utilizing a leader-follower-based consensus algorithm [192]. Since the weighted average value of all DERs' voltages $\sum_{i=1}^{N} \mu_i v_i$ can be forced to converge to it's desired value v^{DES}, then the global voltage regulation can be guaranteed.

Voltage Observer Design

In order to obtain estimate voltage $\hat{v}_i$, the voltage observer is designed as

$$\hat{v}_i(t) = v_i(t_0) + \int_{t_0}^{t} \mathrm{sat}_{\delta_v}\left[\dot{v}_i(\tau) + \sum_{j\in\mathcal{N}_i^v} a_{ij}^v(\hat{v}_j(\tau) - \hat{v}_i(\tau))\right] \mathrm{d}\tau \tag{3.21}$$

where $A^v = (a_{ij}^v)_{N\times N}$ is the adjacency matrix in graph G^v, which represent the communication network of voltage.

We now establish the following theorem.

Theorem 3.2 *If the communication graph of voltage G^v is strongly connected, the designed voltage observer can steer the estimated voltage $\hat{v}_i(t)$ converge to the weighted average value of all DGs' voltage $\sum_{i=1}^{B} \mu_i v_i(t)$.*

Proof Differentiating (3.21), yields

$$\dot{\hat{v}}_i(t) = \mathrm{sat}_{\delta_v}\left[\dot{v}_i + \sum_{j\in\mathcal{N}_i^v} a_{ij}^v(\hat{v}_j(t) - \hat{v}_i(t))\right] \tag{3.22}$$

Denote $\hat{v} = (\hat{v}_1, \cdots, \hat{v}_N)^T$ and $v = (v_1, \cdots, v_N)^T$, the dynamic of the designed voltage observer can be formulated by

$$\dot{\hat{v}}(t) = \phi_{\delta_v}\left[\dot{v}(t) - L^v\hat{v}(t)\right] \tag{3.23}$$

where $L^{v} = (l_{ij}^{v})_{N\times N}$ is the Laplacian matrix and corresponding to the adjacency matrix A^{v}, $\phi_{\delta_v}[\dot{v}_i + \sum_{j\in\mathcal{N}_i^v} a_{ij}^{v}(\hat{v}_j(t) - \hat{v}_i(t))] = (z_1, \cdots, z_n)^T$ with $z_i = \mathrm{sat}_{\delta_v}(\dot{v}_i - \sum_{j\in\mathcal{N}_i^v} \ell_{ij}^{v})$

Then,a set $M = \{\hat{v} : -\delta_v \otimes 1_N \leq \dot{v} - L^v\hat{v} \leq \delta_v \otimes 1_N\}$ can be obtained, which is positive invariant for system (3.23). For the set M, any trajectories of system (3.23) starting outside of M will finally enter M.

Taking $\hat{v} \in M$, system (3.23) can be reduced as

$$\dot{\hat{v}}(t) = \dot{v}(t) - L^v\hat{v}(t) \tag{3.24}$$

Rewriting (3.24) in frequency domain, yields

$$\hat{V} = s(sI_N + L^v)^{-1}V(s) \tag{3.25}$$

where $\hat{V}$ and V are the Laplace transforms of $\hat{v}$ and v, respectively.

Due to the communication graph for voltage G^v is strongly connected, L^v is irreducible. By using the Nyquist stability criterion, the stability of the transfer function $s(sI_N + L^v)^{-1}V(s)$ can be demonstrated. Furthermore, $\sum_{i=1}^{N}$ is an invariant quantity for the positive left eigenvector μ corresponding to the zero eigenvalue of irreducible matrix L^v, the following equality can be obtained through the final value theorem

$$\lim_{t\to+\infty} \hat{v}_i(t) = \lim_{t\to+\infty} \sum_{i=1}^{N} \mu_i v_i(t), i = 1, 2, \cdots, N \tag{3.26}$$

Therefore, the the weighted average value of all DERs' voltages can be asymptotically estimated by the designed voltage observer.

Remark 3.5 The normalized weighted vector μ is determined by the Laplacian matrix L^v in graph G^v, which represent the communication network of voltage. Conversely, if μ is specified in advance, then the corresponding graph G^v needs to be selected. For example, if $\mu = (1/2, 1/3, 1/6)^T$ with $N = 3$, then the strongly connected graph G^v must be selected as an Ring-type digraph with the adjacency matrix A^v only has nonzero weights $a_{13} : a_{32} : a_{21} = 2 : 6 : 3$. As a special case, if G^v is balanced, the normalized weighted vector is selected as $\mu = (1/N, \cdots, 1/N)^T$.

Voltage Regulation and Reactive Power sharing

Across the sparse communication network, the global error of reactive power sharing can be calculated as

$$e_i^Q = \sum_{j\in\mathcal{N}_i^Q} a_{ij}^Q \left(\frac{Q_j}{Q_{j,\max}} - \frac{Q_i}{Q_{i,\max}}\right) \tag{3.27}$$

where $A^Q = (a_{ij}^Q)$ is the adjacency matrix in graph G^Q, which represent the communication network of reactive power.

Then, the distributed cooperative intelligent protocol for voltage regulation and reactive power sharing is designed as

$$\begin{cases} u_i^v = a_{i0}^v(v^{\mathrm{DES}} - \hat{v}_i) + e_i^Q \\ u_i^Q = Q_{i,\max}\mathrm{sat}_{\delta_Q}(e_i^Q) \end{cases} \tag{3.28}$$

where δ_Q is the saturation constant for reactive power control input; $B^v = \mathrm{diag}\{a_{10}^v, \cdots, a_{iN}^v\}$ with a_{i0}^v is the leader adjacency matrix in graph G^v, which represent the communication network of voltage. Assume that there exist one nonzero a_{i0}^v, such that DER_i can receive the reference information v^{DES}.

Then, we can establish the following theorem.

Theorem 3.3 *If the graph G^v and G^P are strongly connected, the voltage regulation and reactive power sharing can be achieved under the proposer distributed intelligent cooperative protocol* (3.28) *with the designed voltage observer* (3.21).

Define the reactive power output ratio is $q_i = Q_i/Q_{i,\max}$. According to (3.21) and (3.28), we have

$$\begin{cases} \dot{\hat{v}}(t) = \mathrm{sat}_{\delta_v}\left[\sum_{j\in\mathcal{N}_i^v}(\hat{v}_j - \hat{v}_i) + a_{i0}^v(v^{\mathrm{DES}} - \hat{v}_i) + e_i^Q\right] \\ \dot{q}_i = \mathrm{sat}_{\delta_Q}(e_i^Q) \end{cases} \tag{3.29}$$

Let $x = -(L^v + B^v)\bar{v} - L^Q q$ and $y = -L^Q q$, where $q = (q_1, \cdots, q_N)^T$, $\bar{v} = (\bar{v}_1, \cdots, \bar{v}_N)^T$ with $\bar{v}_i = \hat{v}_i - v^{\mathrm{DES}}$. Rewrite (3.29) in compact form as

$$\begin{cases} \dot{x} = -(L^v + B^v)\phi_{\delta_v}(x) - L^Q\phi_{\delta_Q}(y) \\ \dot{y} = L^Q\phi_{\delta_Q}(y) \end{cases} \tag{3.30}$$

where $\phi_{\delta_v}(x) = (\mathrm{sat}_{\delta_v}(x_1), \cdots, \mathrm{sat}_{\delta_v}(x_N))^T$, $\phi_{\delta_Q}(x) = (\mathrm{sat}_{\delta_Q}(x_1), \cdots, \mathrm{sat}_{\delta_Q}(x_N))^T$.

Then, (3.30) can be reduced as

$$\dot{x} = -(L^v + B^v)x - L^Q y \tag{3.31}$$

$$\dot{y} = -L^Q y \tag{3.32}$$

To facilitate the proof, we let $a_{ij}^{v} = a_{ij}^{Q}$ and $\delta_v - \delta_Q$. Firstly, we define the following set

$$\begin{aligned}
\mathcal{A}^{+} &= \left\{(x^T, y^T) : x < -\delta_v \otimes 1_N,\ y \geq \delta_v \otimes 1_N\right\} \\
\mathcal{A}^{-} &= \left\{(x^T, y^T) : x < -\delta_v \otimes 1_N,\ 0 \leq y \leq \delta_v \otimes 1_N\right\} \\
\mathcal{B}^{+} &= \left\{(x^T, y^T) : x \geq \delta_v \otimes 1_N,\ y \geq \delta_v \otimes 1_N\right\} \\
\mathcal{B}_1^{-} &= \left\{(x^T, y^T) : 0 \leq x < \delta_v \otimes 1_N,\ y \geq \delta_v \otimes 1_N\right\} \\
\mathcal{B}_2^{-} &= \left\{(x^T, y^T) : \delta_v \otimes 1_N \leq x \leq 0,\ y \geq \delta_v \otimes 1_N\right\} \\
\mathcal{C} &= \left\{(x^T, y^T) : x \geq \delta_v \otimes 1_N,\ 0 \leq y < \delta_v \otimes 1_N\right\} \\
\mathcal{D} &= \left\{(x^T, y^T) : \left|\hat{x}\right| \leq \delta_v \otimes 1_N,\ \left|\hat{y}\right| \leq \delta_v \otimes 1_N\right\}
\end{aligned} \tag{3.33}$$

Let $\mathcal{A} = \mathcal{A}^{+} \cup \mathcal{A}^{-}$, $\mathcal{B} = \mathcal{B}^{+} \cup \mathcal{B}^{-}$ and $\mathcal{B}^{-} = \mathcal{B}_1^{-} \cup \mathcal{B}_2^{-}$. Define $\mathcal{D}_i$, $(i = 1, 2, 3, 4)$ can be represented by

$$\begin{aligned}
\mathcal{D}_1 &= \left\{(x^T, y^T) : x = -\delta_v \otimes 1_N,\ \left|\hat{y}\right| < \delta_v \otimes 1_N\right\} \\
\mathcal{D}_2 &= \left\{(x^T, y^T) : \left|\hat{x}\right| < \delta_v \otimes 1_N,\ y = \delta_v \otimes 1_N\right\} \\
\mathcal{D}_3 &= \left\{(x^T, y^T) : x = \delta_v \otimes 1_N,\ \left|\hat{y}\right| < \delta_v \otimes 1_N\right\} \\
\mathcal{D}_4 &= \left\{(x^T, y^T) : \left|\hat{x}\right| < \delta_v \otimes 1_N,\ y = -\delta_v \otimes 1_N\right\}
\end{aligned} \tag{3.34}$$

Then, restrictions of $\dot{x}$ and $\dot{y}$ on $\partial\mathcal{D}$ are

$$\begin{aligned}
\dot{x}|_{\mathcal{D}_1} &= \left(L^v[\delta_v \otimes 1_N \pm \left|\hat{y}\right|] + B^v \delta_v \otimes 1_N\right) \geq 0 \\
\dot{y}|_{\mathcal{D}_2} &= -L^v \delta_v \otimes 1_N \leq 0 \\
\dot{x}|_{\mathcal{D}_3} &= \left(L^v[\delta_v \otimes 1_N + \left|\hat{y}\right|] + B^v \delta_v \otimes 1_N\right) \leq 0 \\
\dot{y}|_{\mathcal{D}_4} &= -L^v(-\delta_v) \otimes 1_N \geq 0
\end{aligned} \tag{3.35}$$

It follows that no trajectories can leave $\mathcal{D}$ through $\partial\mathcal{D} = \cup_{i=1}^{4} \mathcal{D}_i$, thus, $\mathcal{D}$ is a positively invariant set for (3.30).

And the restrictions of $\dot{x}$ and $\dot{y}$ on sets $\mathcal{A}$, $\mathcal{B}$ and $\mathcal{C}$ can be summarized as follows

$$\begin{aligned}
&\dot{x}|_{\mathcal{A}} = L^{v}(\delta_v \otimes 1_N - |\hat{y}|\,|_{\mathcal{A}}) + B^{v}(\delta_v \otimes 1_N) \geq 0\\
&\dot{y}|_{\mathcal{A}^{+}\cup\mathcal{B}} = -L^{v}(\delta_v \otimes 1_N) \leq 0\\
&\dot{y}|_{\mathcal{A}^{-}\cup\mathcal{C}} = -L^{v}\,|\hat{y}|\,|_{\mathcal{A}^{-}\cup\mathcal{C}} \leq 0\\
&\dot{x}|_{\mathcal{B}^{+}} = -(L^{v} + B^{v})(\delta_v \otimes 1_N) - L^{v}(\delta_v \otimes 1_N) \leq 0\\
&\dot{x}|_{\mathcal{B}_1^{-}} = [-L^{v}(|\hat{x}| + \delta_v \otimes 1_N) - B^{v}\,|\hat{x}|]_{\mathcal{B}^{-}} \leq 0\\
&\dot{x}|_{\mathcal{C}} = -L^{v}(\delta_v \otimes 1_N - |\hat{y}|\,|_{\mathcal{C}}) - B^{v}(\delta_v \otimes 1_N) \leq 0
\end{aligned} \tag{3.36}$$

The restrictions of $\dot{y}$ on the set $\mathcal{B}_2^{-}$ satisfies $\dot{y}|_{\mathcal{B}_2^{-}} \leq 0$ and

$$\begin{aligned}
\dot{x}|_{\mathcal{B}_2^{-}} =& [L^{v}(|\hat{x}| - \delta_v \otimes 1_N) + B^{v}\,|\hat{x}|]|_{\mathcal{B}^{-}}\\
=& \begin{cases} B^{v}\delta_v \otimes 1_N \geq 0, \text{ if } x = -\delta_v \otimes 1_N \\ -L^{v}\delta_v \otimes 1_N \leq 0, \text{ if } x = 0 \end{cases}
\end{aligned} \tag{3.37}$$

which illustrate that all trajectories starting in the set $\mathcal{B}_2^{-}$ will finally enter $\mathcal{D}$.Let $-\mathcal{A} = \{(x^T, y^T) : (-x^T, -y^T) \in \mathcal{A}\}$, $-\mathcal{B} = \{(x^T, y^T) : (-x^T, -y^T) \in \mathcal{B}\}$ and $-\mathcal{C} = \{(x^T, y^T) : (-x^T, -y^T) \in \mathcal{C}\}$. It can be proved by the same method that all the trajectories starting in $-\mathcal{A}$, $-\mathcal{A}$ and $-\mathcal{A}$ will finally enter $\mathcal{D}$.

Thus, the solution of system (3.32) is

$$y(t) = e^{-L^{Q}t}y(0) \tag{3.38}$$

If the graph G^{Q} is strongly connected, it can be obtained from the Theorem 3 in [194] that

$$\lim_{t\to+\infty} y(t) = w_r w_\ell^T y(0) = -w_r w_\ell^T L^{Q} q(0) = 0 \tag{3.39}$$

then, the reactive power ratio of all DERs satisfies $L^{Q}q = 0$, where $q = (q_1, \cdots, q_N)^T$. Therefore, we have $q_i = q_j$ for all $a \neq j$, which deduces the control objective of reactive power sharing is achieved.

In steady state, $y(t) = 0$, the system (3.30) can be reduced as

$$\dot{x} = -(L^{v} + B^{v})x \tag{3.40}$$

Owing to the matrix $L^{v} + B^{v}$ is stable, the asymptotic convergence of system state x can be guaranteed, then

$$\lim_{t\to+\infty} \left|\hat{v}_i - v^{\text{DES}}\right| = 0 \tag{3.41}$$

which implies that the control objective of voltage regulation is achieved.

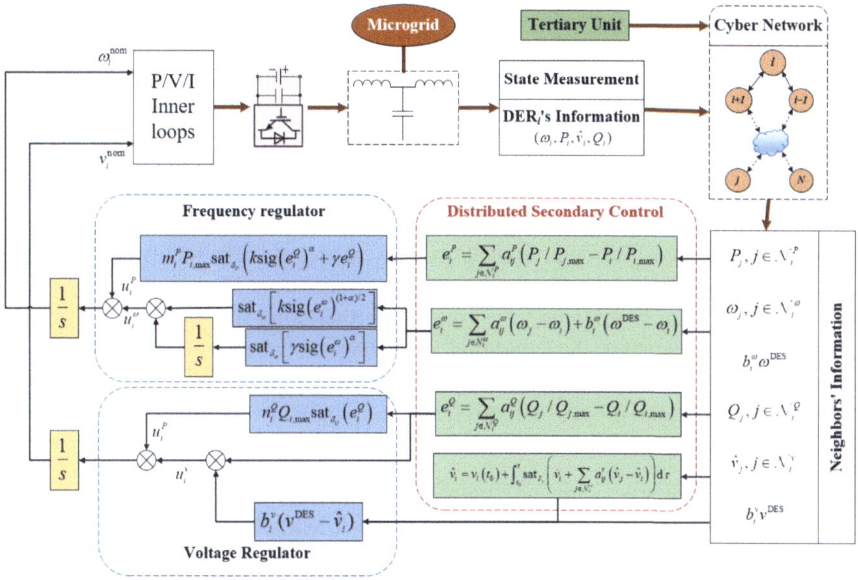

Fig. 3.1 The block diagram of the proposed control scheme. ©2024 IEEE. Reprinted, with permission, from IEEE Transactions on Industrial Informatics [195]

Summarize the above, we conclude that the weighted average value of all DERs' voltage magnitudes can converge to the desired value v^{DES} while maintaining the reactive power sharing under the proposed distributed intelligent cooperative protocols (3.28) with the designed voltage observer (3.21).

According to Theorems 3.1–3.3, the block diagram of proposed distributed intelligent cooperative protocols are depicted as Fig. 3.1. It can be seen that the local controller is implemented by using received measurements from it's neighbors $v_j \in \mathcal{N}_i^v$, $\omega_j \in \mathcal{N}_i^\omega$, $P_j \in \mathcal{N}_i^P$ and $Q_j \in \mathcal{N}_i^Q$, and the reference information from the virtual leader $a_{i0}^v v^{\mathrm{DES}}$ and $a_{i0}^\omega \omega^{\mathrm{DES}}$ through the sparse communication network G^v, G^ω, G^P and G^Q. Then, the calculated global tracking error $\hat{v}_i - v^{\mathrm{DES}}$, e^ω, e^P and e^Q are utilized to regulate nominal set points of each DERs, v_i^{nom} and ω_i^{nom}. By using the designed control framework, the voltage and frequency control process will be endowed with different convergence speed and time scale, which in turn permits us to design these controllers separately. Furthermore, the proposed controller efficiently achieves the contradiction between voltage regulation and reactive power sharing.

Remark 3.6 The asymptotic convergence of the output voltage is rigorously demonstrated by the Lyapunov asymptotic stability analysis. This stability criterion implies that the dynamic trajectories of the output voltage of each DERs will asymptotically converge to a certain bounded domain and stably retain on this domain in the steady state until some new disturbances or external perturbations occur. Therefore, driven by the inherent stability mechanism, all DER's voltages will be effectively forced back to the equilibrium point once they deviate or go out of the acceptable domain under the regulation of the designed voltage controller.

Remark 3.7 In this chapter, we defined four digraphes G^v, G^ω, G^P and G^Q corresponding to communication networks of voltage, frequency, active power and reactive power. The associated weakest connectivity requirements are also determined. This results will benefit the build of appropriate redundant network. Also, the mentioned four communication network can be selected as one uniform strongly connected digraph which contains a spanning tree. Then, a simpler control implementation can be obtained as a result. Furthermore, a connected undirected graph is always satisfies the condition of strongly connection. In spite of this, a sparse communication network can still meet the connectivity requirement of control implement, the proposed controllers are completely distributed and with low communication costs.

Remark 3.8 Compared with the control framework presented in [40], in which each DER must average its own reactive power output with that of all other DERs to complete the update procedure, the proposed controller in this chapter introduces a notable enhancement: it permits the assignment of distinct set points for individual DERs. This capability is crucial, as it explicitly accounts for the non-uniform impedance effects present in distribution lines, which is a factor that often overlooked in conventional designs. If Such non-uniformities are ignored, it will lead to inaccurate reactive power sharing and compromised system stability. By allowing set points to be tailored per DER, the proposed strategy ensures that line impedance variations are intrinsically compensated, thereby improving both accuracy and resilience in reactive power management. Moreover, unlike the methodologies discussed in [40] and [77], the proposed controllers can achieve finite-time frequency restoration even under sparse communication networks with nonidentical edge weights. This is a significant advancement, as sparse communication topologies reduce implementation costs and enhance system robustness against link failures. The incorporation of heterogeneous weights further reflects real-world scenarios where communication channels may differ in reliability or bandwidth. As a result, the control scheme not only accelerates the convergence of frequency to its nominal value but also maintains this performance under more realistic and challenging communication constraints.

Remark 3.9 In contrast to the approach in [196], which assumes that every DER has direct access to the reference voltage, the controllers proposed in this chapter relax this requirement substantially. Specifically, they enable only a small subset of DERs, or even a single DER can obtain reference values. This feature greatly enhances the flexibility and scalability of the control system, particularly in large scale microgrids, where full connectivity to reference signals may be impractical or costly to maintain.

Remark 3.10 The proposed method addresses a critical issue that has been largely neglected in prior finite-time control strategies such as those in [77]: the inherent trade off between accurate voltage regulation and proportional reactive power sharing. While previous studies often prioritize one objective at the expense of the other, our solution incorporates a coordinated mechanism that simultaneously satisfies both requirements. This is achieved through a novel formulation that dynamically adjusts control efforts to balance voltage maintenance with power sharing accuracy, thereby eliminating the need for compromising either aspect.

3.2 Performance Validation

In order to demonstrate the effectiveness of the proposed distributed intelligent cooperative protocol, an islanded microgrid test system consists of 4 DERs, 4 local loads and 4 transmission lines (as shown in Fig. 3.2) is built in MATLAB/SimPowerSystems. The communication network between DERs is depicted in Fig. 3.3 and the detailed system parameters are summarized in Table 3.1. A ring shape communication network with is considered in the islanded microgrid test system. The communication weight of the corresponding unified digraph is 1. Then

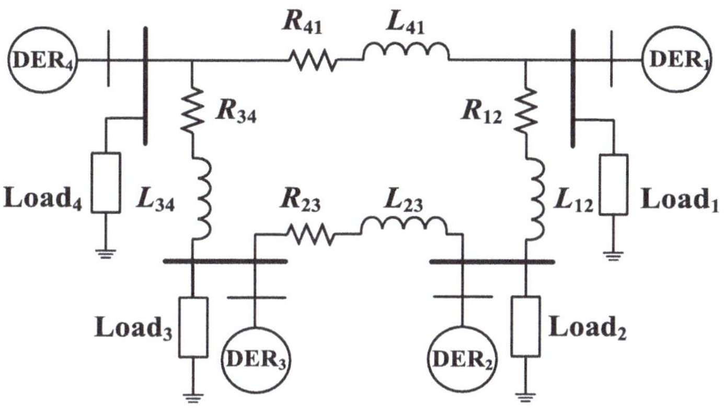

Fig. 3.2 A single line diagram of the microgrid test system. ©2024 IEEE. Reprinted, with permission, from IEEE Transactions on Industrial Informatics [195]

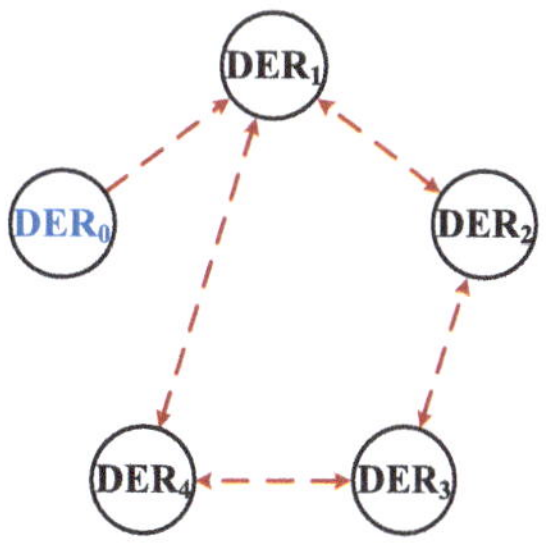

Fig. 3.3 A line type communication network. ©2024 IEEE. Reprinted, with permission, from IEEE Transactions on Industrial Informatics [195]

Table 3.1 The Parameters Values For the Microgrid System

DERs	Parameter	$DER_{1\&3}$	Parameter	$DER_{2\&4}$
	V^{DC}	800 V	V^{DC} V	680 V
	m^P	7.5×10^{-5}	m^P	5×10^{-5}
	n^Q	9×10^{-4}	n^Q	13.5×10^{-4}
	L^f	2 mH	L^f	2 mH
	C^f	16 μF	C^f	16 μF
	L^o	0.21 mH	L^o	0.175 mH
	K^{PV}	10	K^{PV}	8
	K^{IV}	100	K^{IV}	110
	K^{DV}	$1.5e^{-3}$	K^{DV}	$1.4e^{-3}$
	K^{PC}	5	K^{PC}	4
	K^{IC}	200	K^{IC}	180
Loads	$Load_1$	$Load_2$	$Load_3$	$Load_4$
	P_1 19 kW	P_2 18 kW	P_3 22 kW	P_4 17 kW
	Q_1 19 kVar	Q_2 18 kVar	Q_3 22 kVar	Q_4 17 kVar
Lines	$Line_1$	$Line_2$	$Line_3$	$Line_4$
	R_{12} 0.64 Ω	R_{23} 0.51 Ω	R_{34} 0.58 Ω	R_{41} 0.61 Ω
	L_{12} 1.32 mH	L_{23} 1.05 mH	L_{34} 1.21 mH	L_{41} 1.27 mH

the associated adjacency matrices are

$$A^{\upsilon} = A^{\omega} = A^{P} = A^{Q} = \begin{bmatrix} 0\,1\,0\,1 \\ 1\,0\,1\,0 \\ 0\,1\,0\,1 \\ 1\,0\,1\,0 \end{bmatrix} \tag{3.42}$$

The reference values of voltage and frequency is defined as $\upsilon^{\text{DES}} = 380$ V and $\omega^{\text{DES}} = 314$ rad/s, which is generated by a virtual leader and can only be accessed by DER_1. Therefore, the leader adjacency matrix is

$$B^{\upsilon} = B^{\omega} = \begin{bmatrix} 1\,0\,0\,0 \\ 0\,0\,0\,0 \\ 0\,0\,0\,0 \\ 0\,0\,0\,0 \end{bmatrix} \tag{3.43}$$

The Laplacian matrix L^{υ} corresponding to A^{υ}, thus, the associated voltage weighted vector can be calculated as $\mu = (0.25, 0.25, 0.25, 0.25)^T$. The control parameters are set as $k = 3$, $\gamma = 1$ and $\alpha = 5$. And the saturation input constants are chosen as $\delta_{\upsilon} = 10$, $\delta_{\omega} = 2$ and $\delta_P = \delta_Q = 1.5$.

The control performance of the proposed controller will be tested in the following six scenarios: (1) load variation; (2) plug-and-play capability; (3)Communication topology change; (4) communication link failure; (5) communication time delay and (6) data droop. To further illustrate the superiority of the proposed distributed secondary control (PDSC) scheme, some comparison results of other methods including the general asymptotic distributed secondary control (ADSC) scheme and the traditional asymptotic centralized secondary control (ACSC) scheme will be given under the same scenario.

3.2.1 Load Variation

To extensively evaluate the system performance, the effectiveness and dynamic robustness of the proposed controller for handling sudden load changes will be comprehensively verified and analyzed in the following representative scenarios. In the detailed simulation setup, the microgrid test system, which is fully equipped with the PDSC scheme, is initially configured to be operated in the autonomous islanded mode starting at the time instant $t = 0\,\text{s}$ to demonstrate the regulation capability under various conditions.

General Performance Assessment

The simulation results in case of load variation is depicted in Figs. 3.4 and 3.5. It can be observed that the output operating frequencies of all DERs are synchronized to

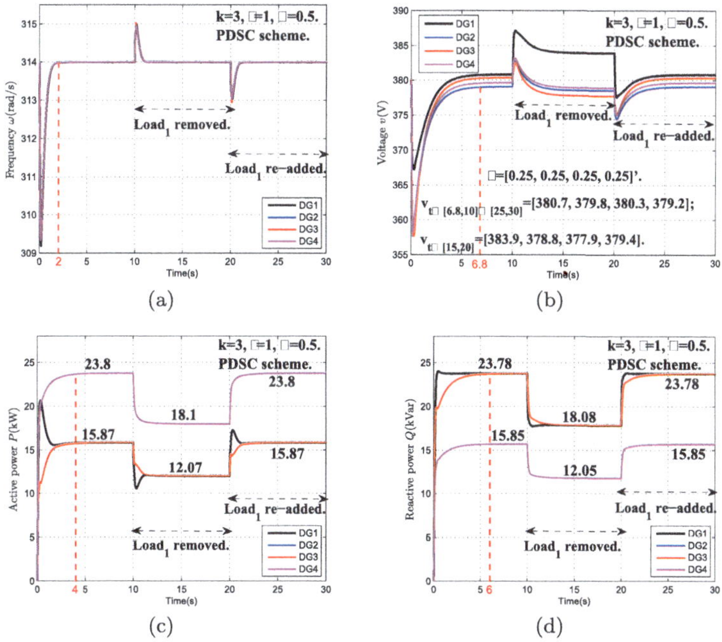

Fig. 3.4 State evolutions for the proposed distributed intelligent cooperative protocol with in case of load change variation: (**a**) operating frequencies; (**b**) voltage amplitudes; (**c**) active power outputs; (**d**) reactive power outputs. ©2024 IEEE. Reprinted, with permission, from IEEE Transactions on Smart Grids [195]

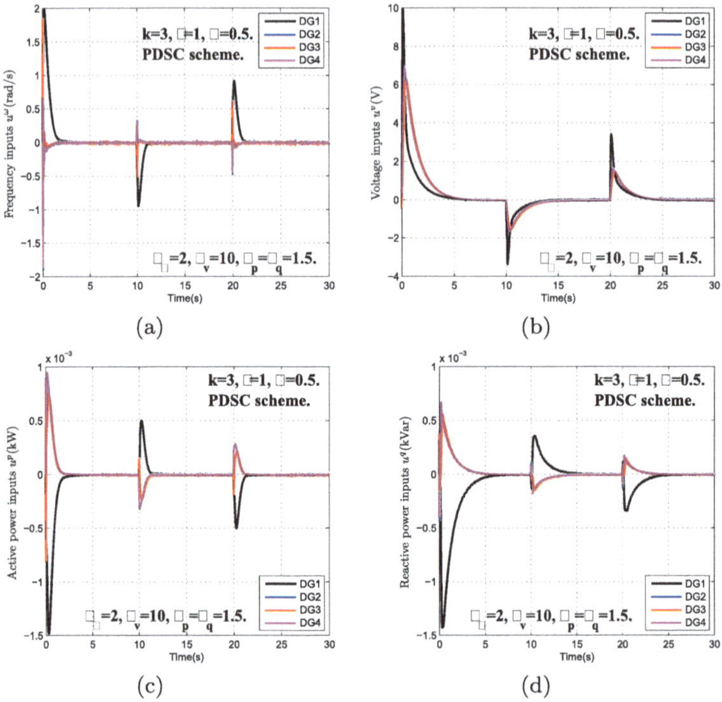

Fig. 3.5 Input evolutions for the proposed distributed intelligent cooperative protocol with in case of load change variation: (**a**) frequency inputs; (**b**) voltage inputs; (**c**) active power inputs; (**d**) reactive power inputs. ©2024 IEEE. Reprinted, with permission, from IEEE Transactions on Smart Grids [195]

the reference value $\omega^{\mathrm{DES}} = 314$ rad/s within $t = 2$ s while the accurate active power sharing is achieved at $t = 4$ s. The observer based distributed intelligent cooperative protocol regulates output voltages of all DERs begin to stabilize near $v^{\mathrm{DES}} = 380$ V after $t = 6.8$ s and keeps accurate reactive power sharing after $t = 6$ s. The local load_1 (19 kW+j19 kVar) is removed at $t = 10$ s and re-added at $t = 20$ s. According to Fig. 3.4, operating frequencies of all DERs can still be rapidly synchronized to the reference value $\omega^{\mathrm{DES}} = 314$ rad/s with the similar transient response after the load change. And output voltages of all DERs are still keep in the acceptable domain. When the load_1 is removed at $t = 10$ s, the rest DERs will implement a new active power sharing scheme. And the active power sharing mode will come back to the previous situation after the load_1 is re-added after $t = 20$ s. The control inputs of voltage, frequency, active and reactive power sharing are shown in Fig. 3.5. Figures 3.6, 3.7, and 3.8 show the control performance of the microgrid test system under PDSC scheme with different control parameters k, γ and α. It can be seen that different values of the control parameters will result in various control performance and convergence speed. We also simulate a microgrid system with 8 DG in Fig. 3.9 to test the control performance of PDSC scheme in large size microgrid. As seen,

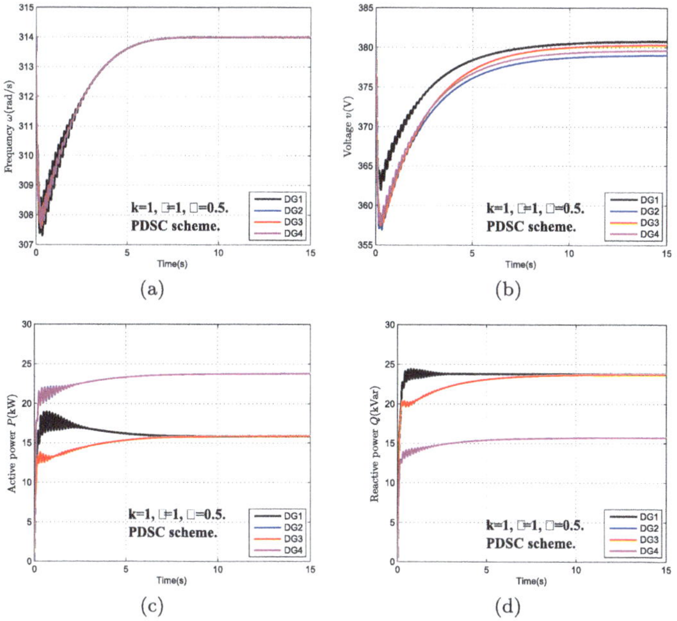

Fig. 3.6 State evolutions for the proposed distributed intelligent cooperative protocol with different control parameters: (**a**) operating frequencies; (**b**) voltage amplitudes; (**c**) active power outputs; (**d**) reactive power outputs. ©2024 IEEE. Reprinted, with permission, from IEEE Transactions on Smart Grids [195]

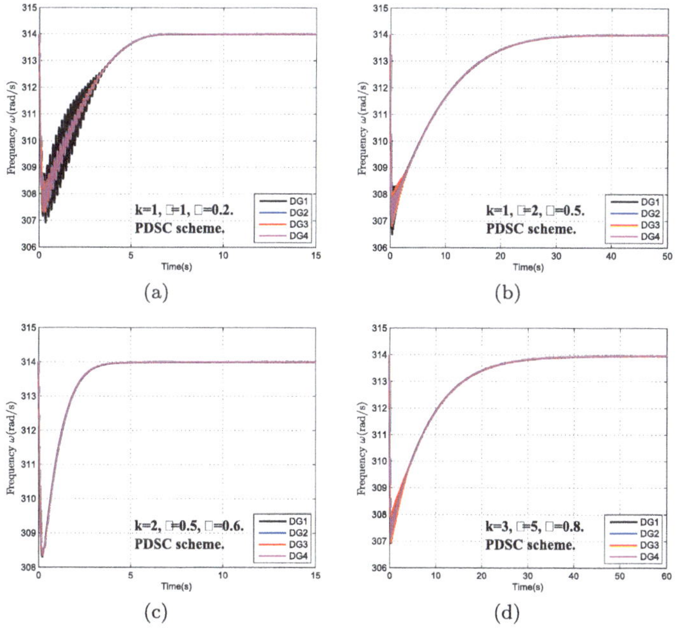

Fig. 3.7 Frequency response for the proposed distributed intelligent cooperative protocol with different control parameters. ©2024 IEEE. Reprinted, with permission, from IEEE Transactions on Smart Grids [195]

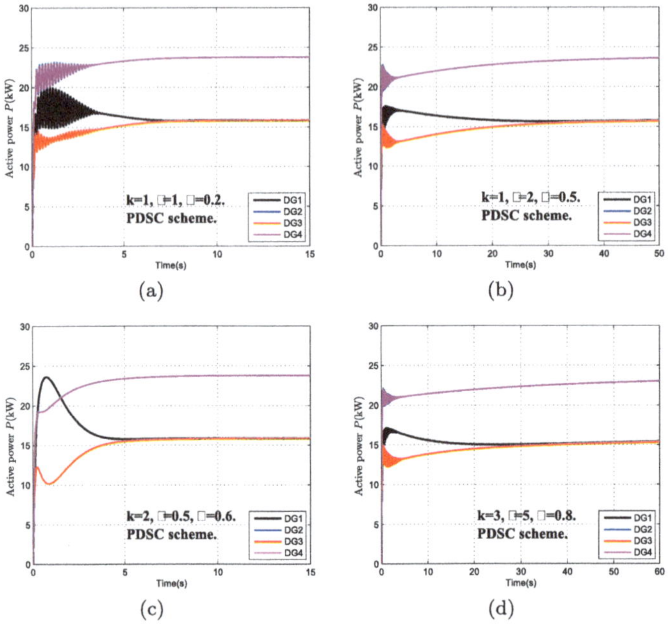

Fig. 3.8 Active power outputs for the proposed distributed intelligent cooperative protocol with different control parameters. ©2024 IEEE. Reprinted, with permission, from IEEE Transactions on Smart Grids [195]

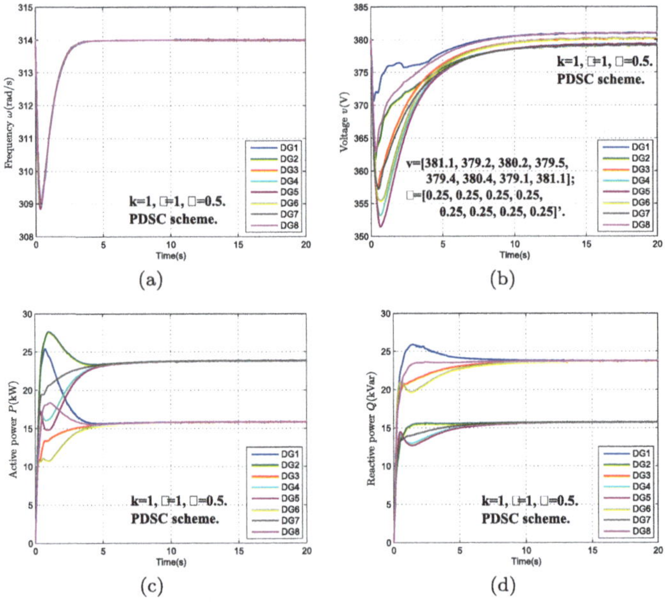

Fig. 3.9 State evolutions for the proposed distributed intelligent cooperative protocol in case of microgrid consists of 8 DERs and 8 local loads, where the physics and cyber networks are ring shape: (**a**) operating frequencies; (**b**) voltage amplitudes; (**c**) active power outputs; (**d**) reactive power outputs. ©2024 IEEE. Reprinted, with permission, from IEEE Transactions on Smart Grids [195]

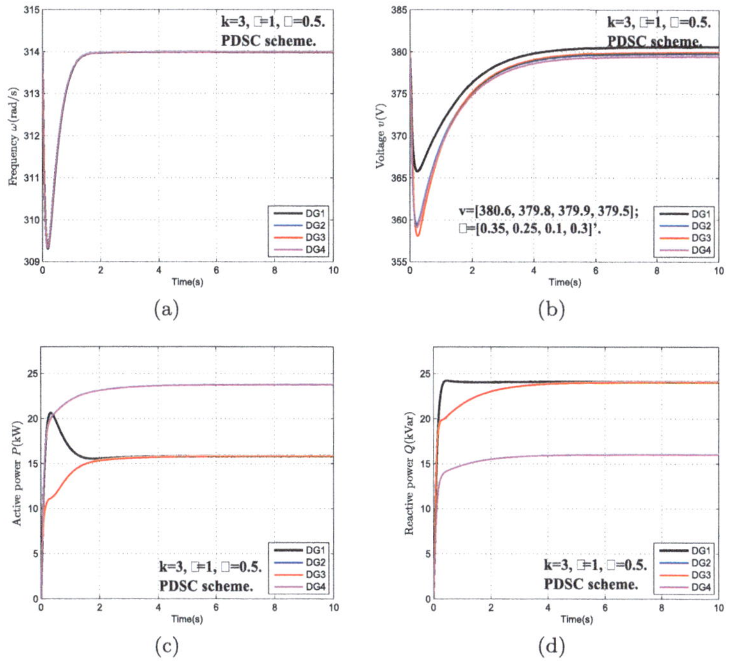

Fig. 3.10 State evolutions for the proposed distributed intelligent cooperative protocol with a pre-specified weighted vector $\mu = (0.35, 0.25, 0.1, 0.3)^T$ which is associated to the adjacency matrix A^v with elements $a_{21} = 1.2$, $a_{32} = 3$, $a_{43} = 1$, $a_{14} = 0.86$ and $a_{ij} = 0$ for other i and j: (**a**) operating frequencies; (**b**) voltage amplitudes; (**c**) active power outputs; (**d**) reactive power outputs.

the voltage regulation, frequency restoration, active and reactive power sharing can be still achieved with longer convergence time.

Test for Voltage Weighted Vector μ

It can be seen from Fig. 3.4b that the weighted average value of all DER's voltage can be calculated as $\mu^T v = 380$ V, for $t \in [6.8, 10]$ s. And Fig. 3.10 shows the state evolution under the proposed PDSC scheme with the pre-specified weighted vector $\mu = (0.35, 0.25, 0.1, 0.3)^T$. As seen, the weighted average value of all DER's voltage can be calculated as $\mu^T v = 380$ V, which implies that the voltage control objective can be achieved under PDSC scheme with a directed communication network designed by the weighted vector μ.

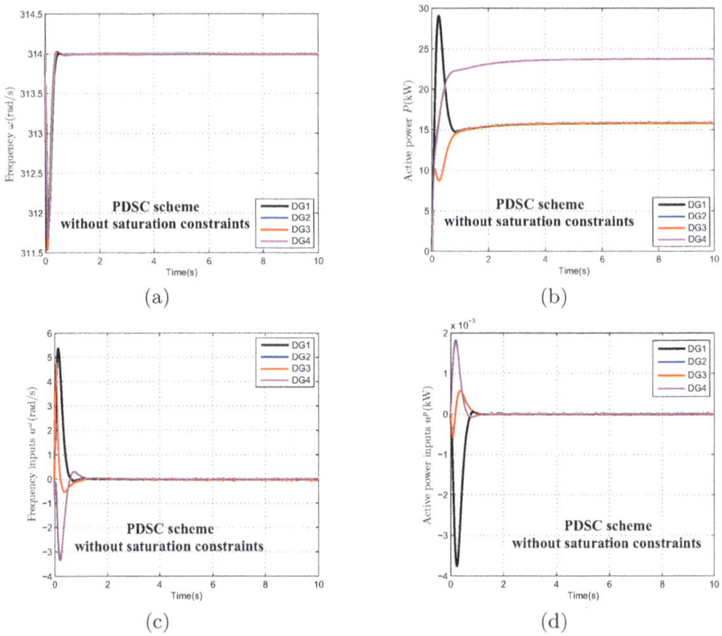

Fig. 3.11 State and control input evolutions for the proposed distributed intelligent cooperative protocol without the saturation constraint: (**a**) operating frequencies; (**b**) active power outputs; (**c**) frequency inputs; (**d**) active power inputs. ©2024 IEEE. Reprinted, with permission, from IEEE Transactions on Smart Grids [195]

Boundedness of the Control Inputs

The bounded control inputs of voltage, frequency, active and reactive power are depicted in Fig. 3.5, respectively. For comparison, the control performance under PDCS with unbounded control inputs, as shown in Fig. 3.11a, and b, and the associated control inputs are depicted in Fig. 3.11c and d. It can be seen that there are more transient overshoot for the frequency and active power control.

ADSC Scheme With or Without Voltage Observer

We also simulate the PDSC scheme with or without the voltage observer by setting $\alpha = 1$. The corresponding simulation results are depicted in Figs. 3.12 and 3.13, respectively. In Fig. 3.12, due to the inherent contradiction between voltage regulation and reactive power sharing, the accurate reactive power sharing of test microgrid system under PDSC scheme without using the voltage observer cannot be achieved. Furthermore, it can be seen from Fig. 3.13 that convergence time of voltage, frequency, active and reactive power are 8 s, 10 s, 11 s and 9 s,

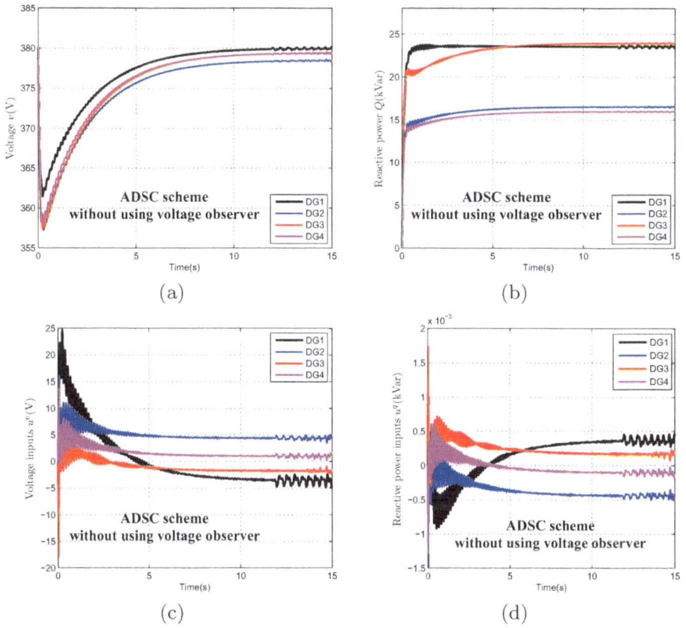

Fig. 3.12 State and control input evolutions for the general ADSC scheme without the voltage observer: (**a**) voltage amplitudes; (**b**) reactive power outputs; (**c**) voltage inputs; (**d**) reactive power inputs. ©2024 IEEE. Reprinted, with permission, from IEEE Transactions on Smart Grids [195]

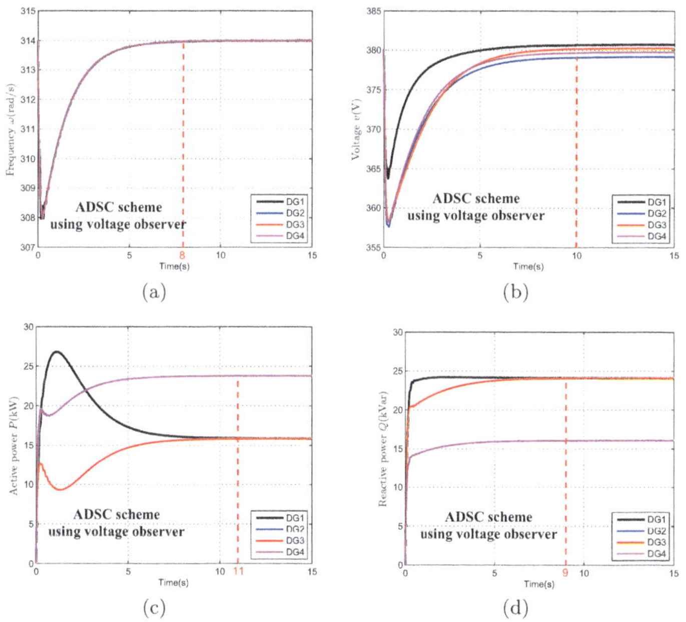

Fig. 3.13 State evolutions for the general ADSC scheme with the voltage observer: (**a**) operating frequencies; (**b**) voltage amplitudes; (**c**) active power outputs; (**d**) reactive power outputs. ©2024 IEEE. Reprinted, with permission, from IEEE Transactions on Smart Grids [195]

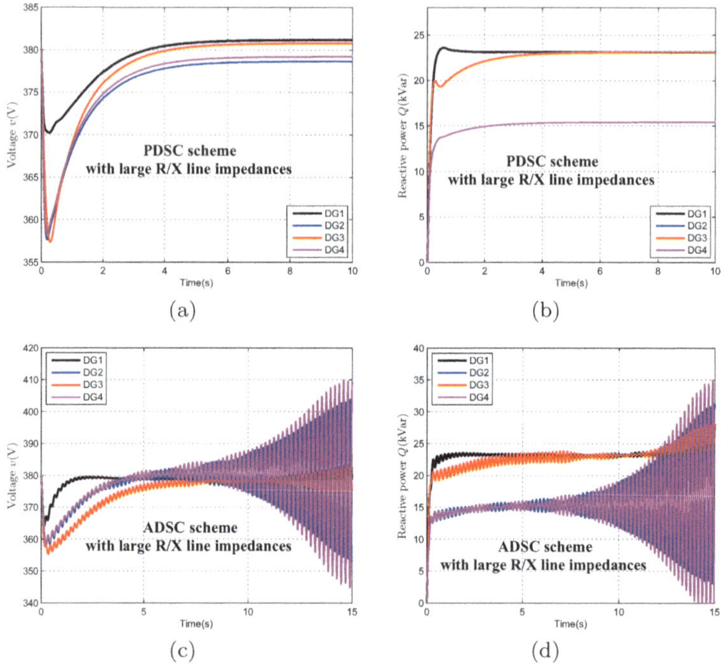

(a) (b) (c) (d)

Fig. 3.14 State evolutions for different control scheme with large R/X line impedances: (**a**) voltage amplitudes under the PDSC scheme; (**b**) reactive power outputs under the PDSC scheme; (**c**) voltage amplitudes under the ADSC scheme; (**d**) reactive power outputs under the ADSC scheme. ©2024 IEEE. Reprinted, with permission, from IEEE Transactions on Smart Grids [195]

respectively. Comparing those in Fig. 3.4, we can conclude that the PDSC scheme has mode rapid convergence time than the general ADSC scheme.

PDSC Scheme vs ADSC With Large R/X Line Impedance

The PDSC and the general ADSC schemes are simulated in microgrid test system with large R/X line impedances. The corresponding simulation results are depicted in Fig. 3.14. The four line impedances RL line$_{1-4}$, as shown in Fig. 3.2 are reset as $0.96 + j0.41\ \Omega$, $0.76 + j0.33\ \Omega$, $0.87 + j0.38\ \Omega$ and $0.91 + j0.40\ \Omega$, respectively. The proposed controller with finite time regulator leads to a faster convergence time of frequency restoration and active power sharing, which effectively distinguish the convergence time of voltage restoration and reactive power sharing in a slower time scale. The designed control framework decouples the voltage and frequency control to a certain extent, which further optimizes the control performance of the PDSC scheme. Therefore, it can be observed from Fig. 3.14 that the PDSC scheme has better control performance than the general ADSC scheme.

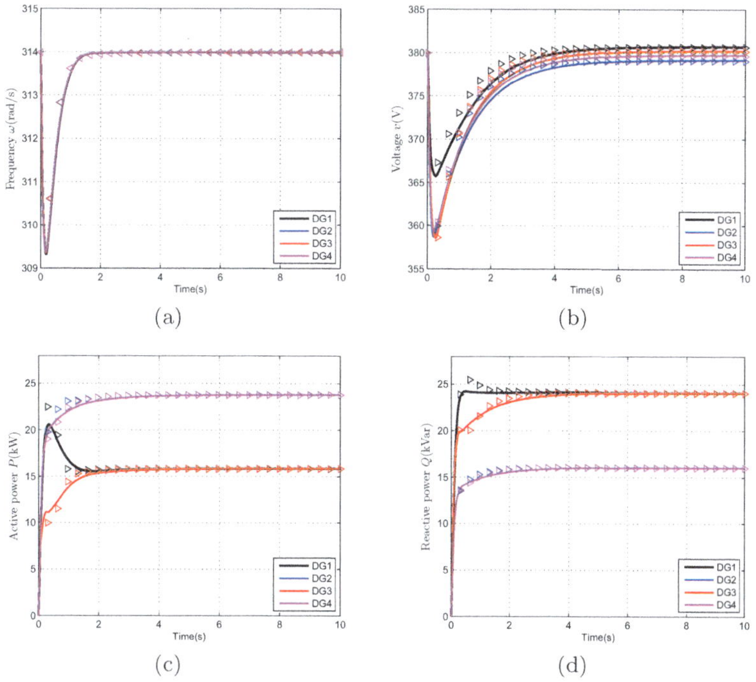

Fig. 3.15 Step response of the control system: (**a**) operating frequencies; (**b**) voltage amplitudes; (**c**) active power outputs; (**d**) reactive power outputs. ©2024 IEEE. Reprinted, with permission, from IEEE Transactions on Smart Grids [195]

Step Response Test

To further validate the control performance of the PDSC scheme, the step response of the microgrid test system is simulated. As shown in Fig. 3.15, all the step response curves fit well.

3.2.2 *Plug-and-Play Capability*

The plug-and-play capability of the PDSC scheme is analyzed in this section. The microgrid test system consists of 4 DERs is operated at $t = 0\,\text{s}$. The DER_5 (with the same parameters as DER_2 and the associated line impedance is $0.62 + j0.4\ \omega$) is plugged in the microgrid at $t = 5\,\text{s}$ and DER_4 is plugged out from the microgrid at $t = 10\,\text{s}$. For comparison, the microgrid test system under the general ADSC scheme is simulated with the DER5 is plugged in the microgrid at $t = 12\,\text{s}$ and DER_4 is plugged out from the microgrid at $t = 21\,\text{s}$. The communication weights, as shown in Fig. 3.2 is increased to 2. The corresponding simulation results are depicted in Figs. 3.16 and 3.17.

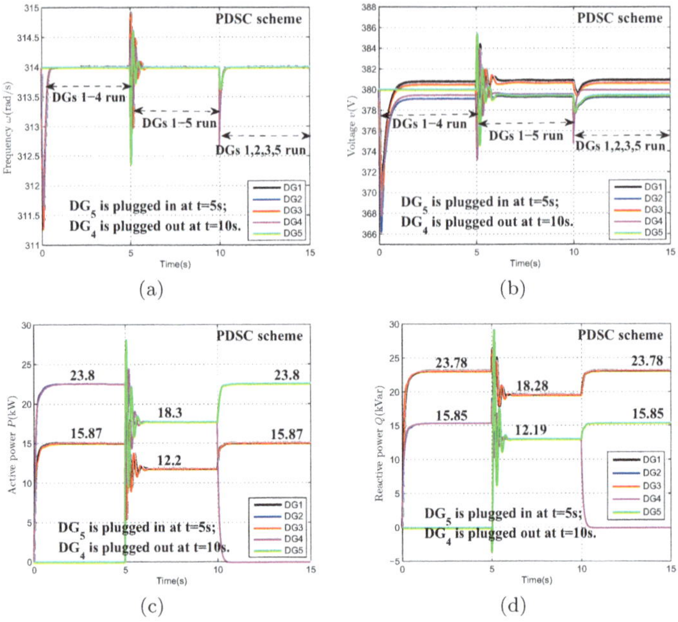

Fig. 3.16 State and control input evolutions for PDSC scheme in case of DER plug and play operation: (**a**) operating frequencies; (**b**) voltage amplitudes; (**c**) active power outputs; (**d**) reactive power outputs. ©2024 IEEE. Reprinted, with permission, from IEEE Transactions on Smart Grids [195]

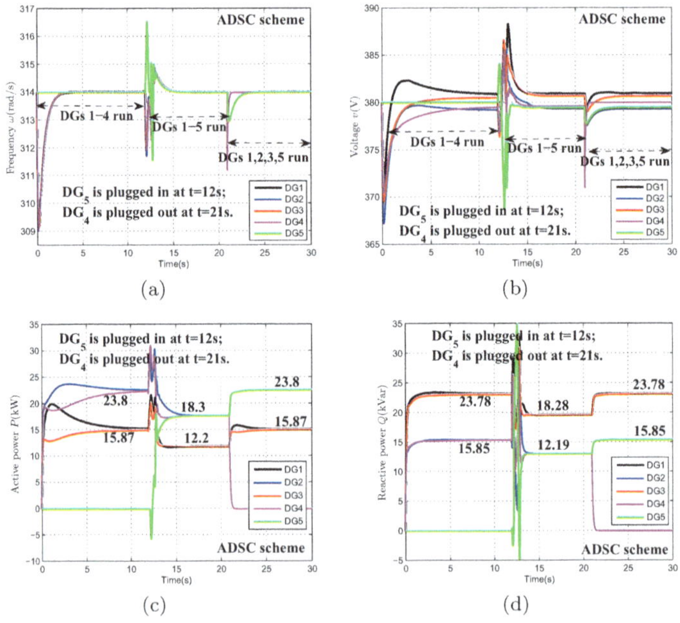

Fig. 3.17 State and control input evolutions for ADSC scheme in case of DER plug and play operation: (**a**) operating frequencies; (**b**) voltage amplitudes; (**c**) active power outputs; (**d**) reactive power outputs. ©2024 IEEE. Reprinted, with permission, from IEEE Transactions on Smart Grids [195]

With the addition of DER_5, the communication network between DER_1 and DER_5 is constructed with weight 2. It can be seen from Fig. 3.16, the voltage regulation, frequency restoration, active and reactive power sharing can be achieved with in 3 s after DER_5 is plugged in at $t = 5$ s. When DER_4 is plugged out at $t = 10$ s, the output power of DER_4 will rapidly decrease to zero, and the rest DERs will generate more power to realize a new power sharing mode. The excision of DER_4 implies loss of the communication link. However, the rest communication links can still form a connected digraph, thus, the control system is still available.

Compare with the control performance of the microgrid test system under the general ADSC scheme shown in Fig. 3.17, the PDSC scheme possesses mode rapid convergence speed and more excellent steady-state performance.

3.2.3 Communication Topology Change

The control performance of the PDSC scheme under four communication topologies including full connected, star shape, tree shape and line shape is analyzed in this section. Compare the simulation results, as shown in Figs. 3.18, 3.19, 3.20, and 3.21 and those in Figs. 3.4 and 3.5, we conclude that there are same convergence time of the PDSC scheme under full connected and faster than that under other topologies. The convergence time of the PDSC scheme under ring shape communication network is faster than that under tree shape and line shape topologies. And the line shape communication network has the slowest convergence time. This is because that the convergence speed is determined by the coefficient matrix in control system, which is affected by the the communication network. The convergence time of the PDSC scheme under four communication topologies is depicted in Table 3.2. It can be seen that there is no significant differences in the frequency and active power convergence times of the PDSC scheme under four communication topologies. This is because of the finite time controller design. However, the convergence times of voltage and reactive power under four communication topologies is have obvious differences due to the designed controller with the asymptotic convergence characteristic.

3.2.4 Link Failure Test

The robustness of the PDSC scheme to the communication failure is tested by comparing with the traditional ADSC and ACSC schemes. The communication links between DER_1 and DER_2, and DER_1 and DER_4 are randomly disconnected at $t \in [0.5, 2]$ s and $t \in [3, 4.5]$ s, respectively. Furthermore, the communication link between the virtual leader DER_0 and DER_1 is disconnected at $t \in [5.5, 7]$ s. The simulation results of the PDSC scheme is depicted in Fig. 3.22(a1)–(a4). The micro-grid test system under the PDSC scheme is almost free from the communication link

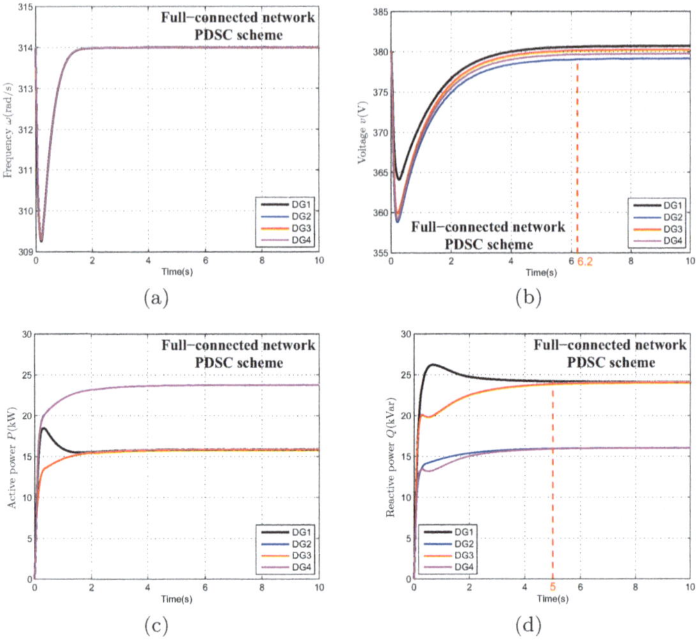

Fig. 3.18 State evolutions for the PDSC scheme under full-connected network: (**a**) operating frequencies; (**b**) voltage amplitudes; (**c**) active power outputs; (**d**) reactive power outputs. ©2024 IEEE. Reprinted, with permission, from IEEE Transactions on Smart Grids [195]

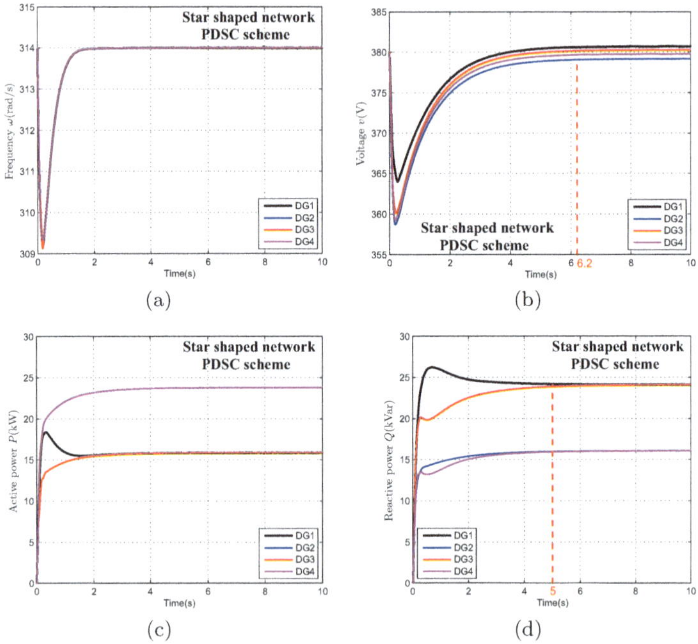

Fig. 3.19 State evolutions for the PDSC scheme under star shaped network: (**a**) operating frequencies; (**b**) voltage amplitudes; (**c**) active power outputs; (**d**) reactive power outputs. ©2024 IEEE. Reprinted, with permission, from IEEE Transactions on Smart Grids [195]

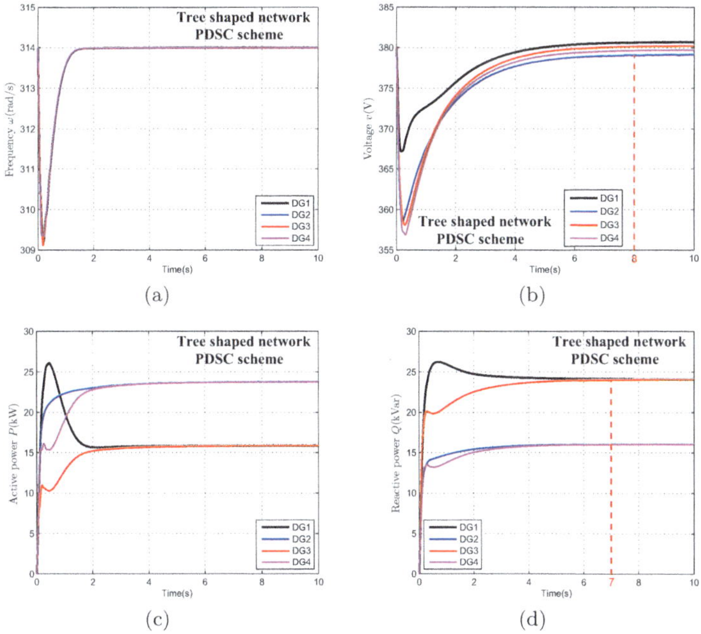

Fig. 3.20 State evolutions for the PDSC scheme under tree shaped network: (**a**) operating frequencies; (**b**) voltage amplitudes; (**c**) active power outputs; (**d**) reactive power outputs.

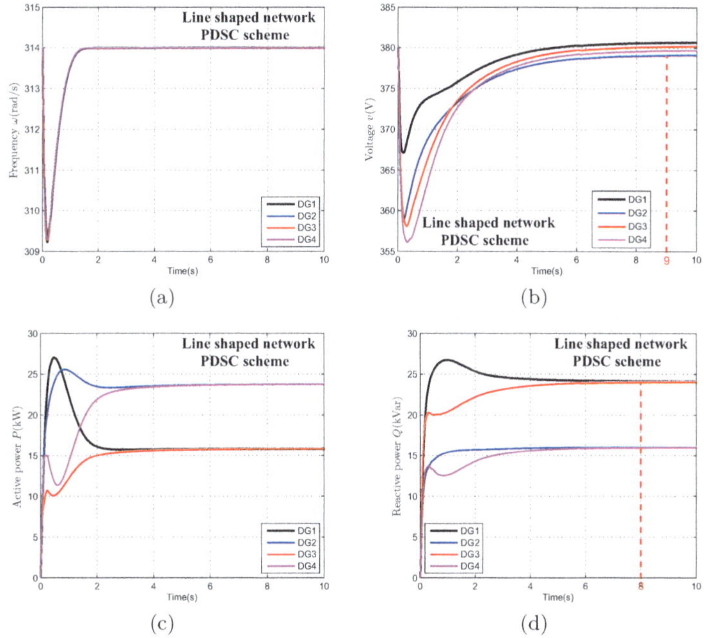

Fig. 3.21 State evolutions for the PDSC scheme under line shaped network: (**a**) operating frequencies; (**b**) voltage amplitudes; (**c**) active power outputs; (**d**) reactive power outputs.

Table 3.2 The parameters for different communication topologies

Different topos		Full	Star	Ring	Tree	Line
Spectral radius		0.2087	0.2087	0.1864	0.1392	0.1206
Convergence time	ω	1.9 s	1.9 s	2 s	2.1 s	2.2 s
	P	3 s	3 s	4 s	4.5 s	4.5 s
	Q	5 s	5 s	6 s	7 s	8 s
	v	6.2 s	6.2 s	6.9 s	8 s	9 s

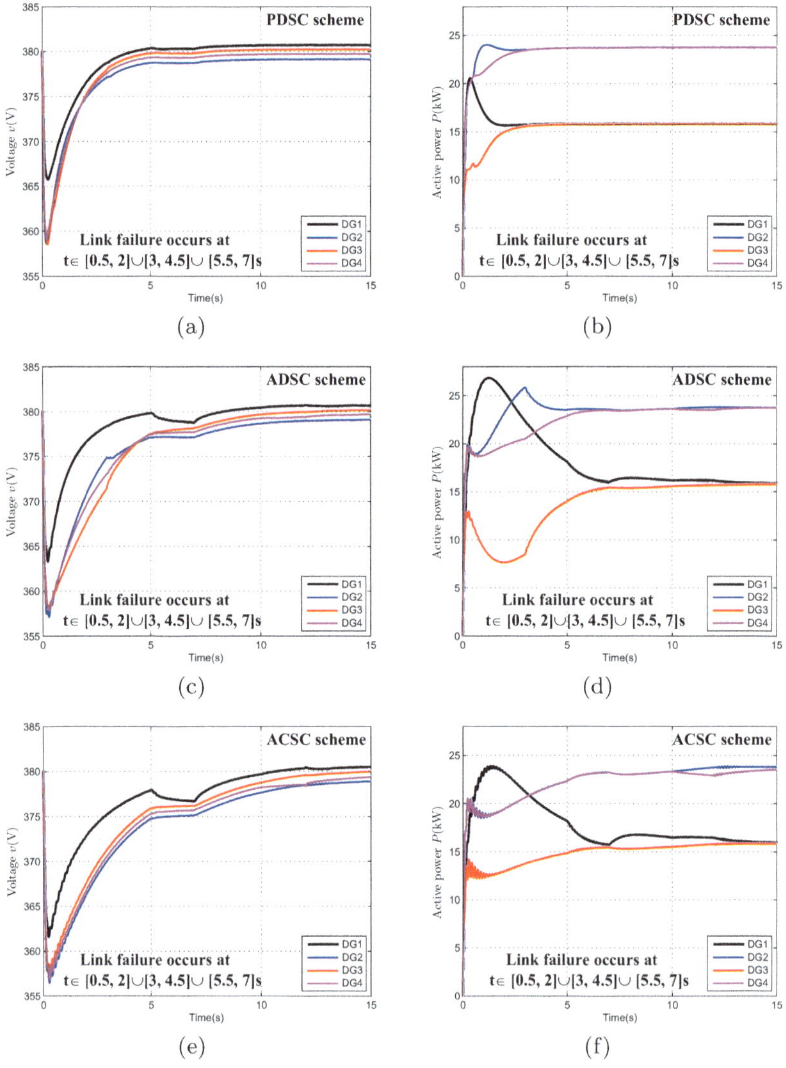

Fig. 3.22 State evolutions under different control scheme in case of the communication link failure: (**a**), (**c**) and (**e**): voltage amplitudes; (**b**), (**d**), and (**f**): reactive power outputs. ©2024 IEEE. Reprinted, with permission, from IEEE Transactions on Smart Grids [195]

failure between the virtual leader DER_0 and DER_1, the voltage regulation, frequency restoration, active and reactive power sharing can still be achieved. It can be seen that the PDSC scheme has excellent robustness to the communication failure. The ACSC scheme has rapider convergence speed than the ADSC scheme. However, the ADSC scheme has better transient control performance. Therefore, based on the robustness of the PDSC scheme, we can timely break some communication links to ensure the security of information to deal with the secure communication problem.

3.2.5 Robustness to Communication Time Delays

The impact of communication time delays on the PDSC scheme is tested here by comparing with the traditional ADSC scheme. The corresponding simulation results are depicted in Figs. 3.23 and 3.24. Both the PDSC scheme and the traditional ADSC scheme show robustness to small communication time delays. However, it can be seen that the PDSC scheme shows better robustness to communication

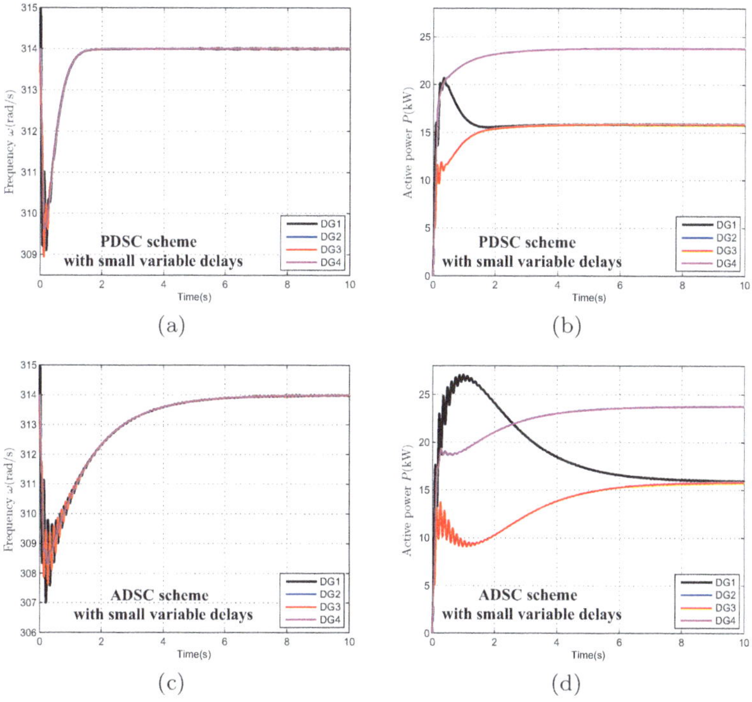

Fig. 3.23 State evolutions under different control schemes in case of the communication time delay $0.08\sin(10t + \pi/4) + 0.09$: (**a**) and (**c**): operating frequencies; (**b**) and (**d**): active power outputs. ©2024 IEEE. Reprinted, with permission, from IEEE Transactions on Smart Grids [195]

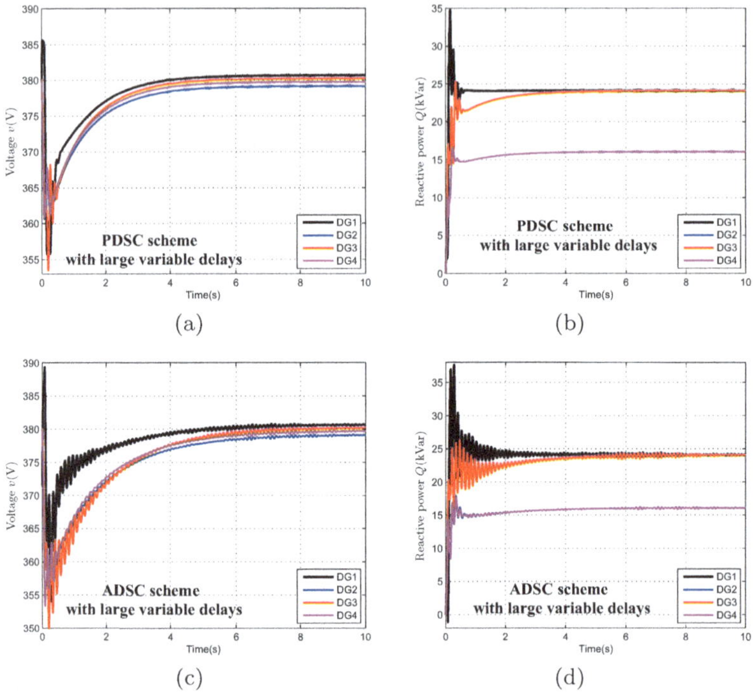

Fig. 3.24 State evolutions under different control schemes in case of the large communication time delay $0.1 \sin(10t + \pi/4) + 0.15$: (**a**) and (**c**): voltage amplitudes; (**b**) and (**d**): reactive power outputs. ©2024 IEEE. Reprinted, with permission, from IEEE Transactions on Smart Grids [195]

time delays. Therefore, the finite time controller possess more robustness to communication time delays.

3.2.6 *Impact of Data Droop-out*

The impact of data droop-out on the PDSC scheme is tested in two scenarios: (1) the data droop-out is occurs once in 10 ms with the consideration of communication time delay is set as 2.5 ms; (2) the data droop-out is occurs once in 20 ms with the consideration of communication time delay is set as 8 ms. The corresponding simulation results are shown in Figs. 3.25 and 3.26, respectively. And the associated control inputs of the PDSC scheme are depicted in Fig. 3.27.

It can be seen that both the PDSC scheme and the ADSC scheme have acceptable control performance in scenario (1) While the PDSC scheme shows better control performance in scenario (2) The robustness of the PDSC to the randomly packet loss is also tested and the corresponding results are depicted in Fig. 3.28. The control objectives can be achieved at $t = 10$ s and not be significantly affected afterwards.

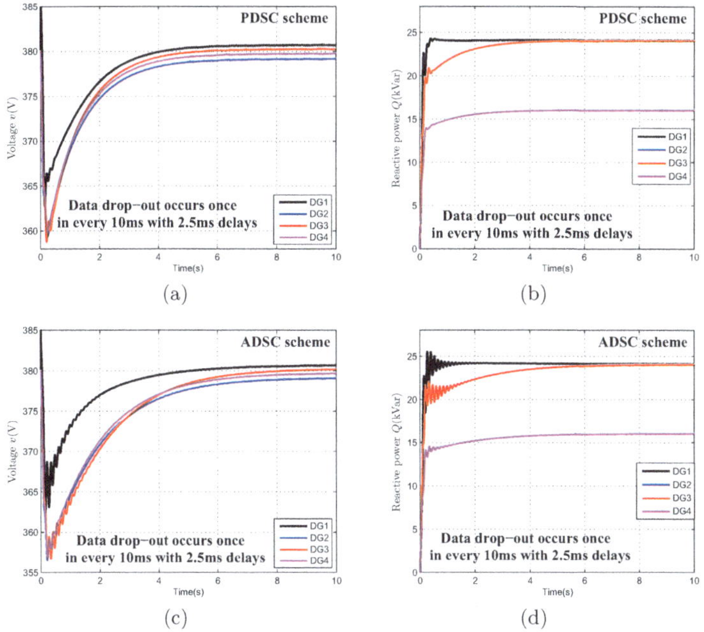

Fig. 3.25 State evolutions under different control schemes in case of the data droop-out in all communication links ar each time: (**a**) and (**c**): voltage amplitudes; (**b**) and (**d**): reactive power outputs. ©2024 IEEE. Reprinted, with permission, from IEEE Transactions on Smart Grids [195]

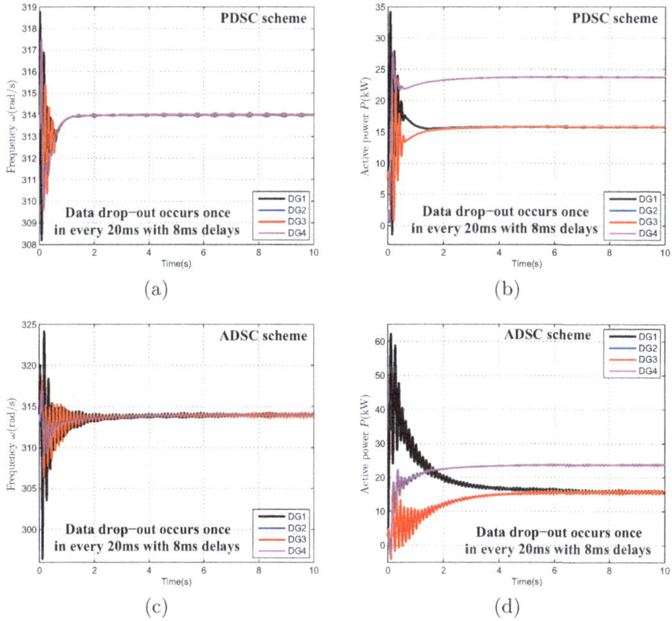

Fig. 3.26 State evolutions under different control schemes in case of the data droop-out in randomly selected communication links ar each time: (**a**) and (**c**): operating frequencies; (**b**) and (**d**): active power outputs. ©2024 IEEE. Reprinted, with permission, from IEEE Transactions on Smart Grids [195]

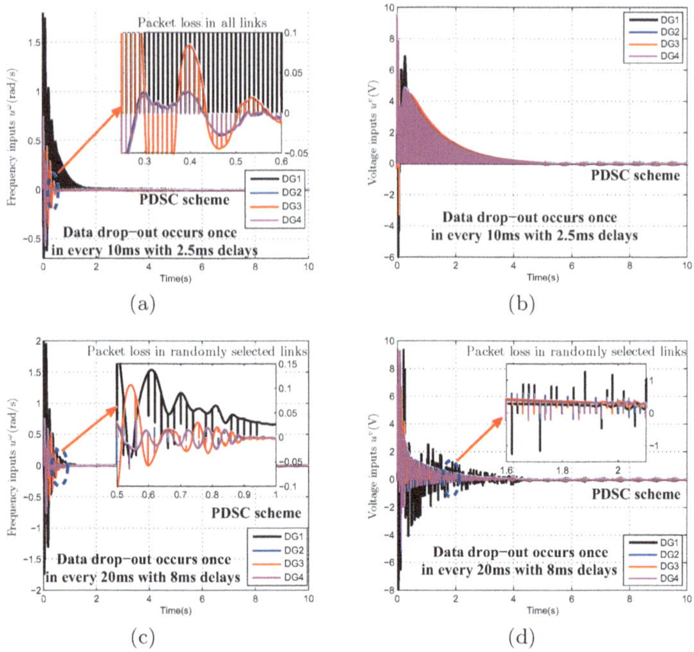

Fig. 3.27 Control input evolutions under the PDSC schemes in case of the data droop-out: (**a**) and (**c**): frequency inputs; (**b**) and (**d**): voltage inputs. ©2024 IEEE. Reprinted, with permission, from IEEE Transactions on Smart Grids [195]

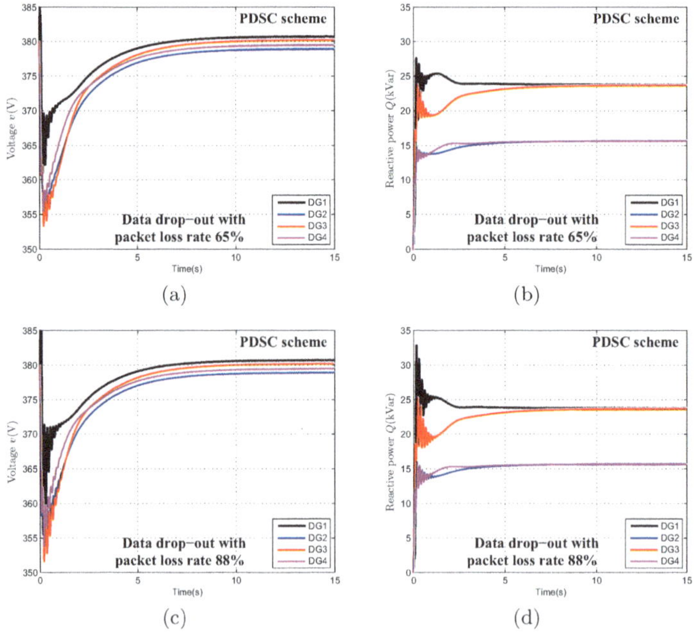

Fig. 3.28 State evolutions under the PDSC schemes in case of the data droop-out: (**a**) and (**c**): voltage amplitudes; (**b**) and (**d**): reactive power outputs. ©2024 IEEE. Reprinted, with permission, from IEEE Transactions on Smart Grids [195]

3.3 Summary

This chapter proposes two distributed intelligent cooperative protocols for weighted average voltage regulation, frequency restoration, active and reactive power sharing in microgrids. Under the proposed controller with finite time regulator, the frequency restoration and accurate active power sharing can be realized within finite time, which ensures the voltage regulation and reactive power sharing to be achieved in a slower time scale. It in turn reduce the inherent coupling of frequency and voltage, and permits us to design these controllers separately. By using the proposed weighted average voltage observer, a compromise can be utilized, which ensures the accurate reactive power sharing can be guaranteed while maintaining the voltage regulation in large R/X microgrids. By using the saturation constraint, all controllers are designed with bounded control inputs, thus, the problem of consequent transient overshoot and the high frequent oscillation near the equilibrium has been greatly suppressed. Finally, the simulation results effectively proved that the proposed distributed cooperative intelligent protocols can regulate the weighted average voltage, restore the frequency and share the active and reactive power in microgrid even in the presence of link failure, communication delays, and data packet losses.

Chapter 4
Robust Distributed Cooperative Intelligence for AC Microgrids with Time Varying Delays and Dynamical Network Topology

This chapter develops a droop-based robust distributed cooperative control scheme for AC microgrids under a switching communication network with non-uniform time varying delays. This chapter illustrates a pinning-based frequency/voltage controllers containing a distributed voltage observer. In addition, a consensus-based active/reactive controllers are applied into the distributed intelligent cooperative protocol to generate the nominal set points used in the primary control stage for distributed energy resources. By employing these approaches, the frequency and weighted average DER agents' voltage can be pinned to a given value and power sharing of active and reactive power is maintained. Finally, by considering the non-uniform delay in the communication network between distributed DER agents, this chapter proves the sufficient conditions of network connectivity and delay upper bound to ensure the reliability and stability of microgrids.

The existing hierarchical control schemes of microgrids mainly include primary control (droop control) [197], secondary control [198] (restoration of voltage and frequency) and tertiary control (optimal dispatch) [199]. The communication structure of the existing secondary control strategies has undergone a development process from centralized control [200, 201] (with low reliability), decentralized control [135, 202, 203] to the current distributed control [192, 193, 204–208] (with high reliability and low cost). This draws on the cooperative control of multi-agent systems [209–212], allowing all DER agents to communicate and share information with their neighbors.

In all previous researches, feedback linearization [191] is widely used to transform the nonlinear dynamics of DER agents to linear dynamics. As a result, the traditional secondary voltage control leads to a first-order tracking synchronization problem. Due to the line impedance effect, the conventional voltage droop control law is unable to share reactive power accurately among even identical DER agents [107, 213]. There is also a new reactive power sharing scheme [40], where each DER agent needs to take the average reactive power value of all DER agents as its own nominal state to complete the update process. As a result, reactive power sharing

J. Lai, X. Lu, *Distributed Cooperative Intelligence in Microgrid Control*,
https://doi.org/10.1007/978-981-95-6884-0_4

can be unaffected by voltage sensing mismatches or line impedance. However, the above reactive power sharing approach cannot guarantee the steady state of the DER agents' voltage due to the contradiction between precise voltage estimation and precise reactive power sharing.Based on the above researches analysis, distributed communication network is the basis for designing microgrids' control strategy. In this case, network variability and communication delay cannot be ignored in the analysis of system stability. In addition, for microgrids with a large number of DER agents, with the introduction of a pinning-based control strategy, it only needs feedback control for a small number of DER agents.

This chapter illustrates the distributed cooperative frequency/voltage control strategy in switching communication networks, and considers the influence of non-uniform time varying delay on active/reactive power sharing. This chapter firstly convert the issue of frequency/voltage regulation to a pinning synchronization problem inspired by the pinning control idea [211, 212, 214]. In this strategy, only a small number of DERs(operate as a leaders) have access to reference values, while the rest are synchronized with the leaders through a distributed communication network. Moreover, this chapter also transform the issue of active/reactive power sharing to an average consensus problem [209, 210]. In the case, Each DER adjusts its output power according to its own output power ratio and that of its neighbors. Accordingly, a distributed cooperative control scheme containing a pinning-based frequency/voltage controller and a consensus-based active/reactive controller are proposed. These controllers can generate the nominal set-points of the droop control so as to realize the frequency restoration and voltage regulation while maintaining the accurate active and reactive power sharing.

4.1 Robust Distributed Cooperative Intelligence

In this section, we illustrate the distributed cooperative control scheme for voltage/frequency control and active/reactive power sharing, respectively. As shown in (4.1), the droop technique applied in primary control can be expressed as [215]

$$\begin{cases} v = v^{\mathrm{nom}} - n^{Q} Q \\ \omega = \omega^{\mathrm{nom}} - m^{P} P \end{cases} \tag{4.1}$$

where v^{nom} and ω^{nom} represents nominal set point of voltage and frequency. n^{Q} and m^{P} are the reactive/active droop coefficients according to the output power ratios, Q and P are the measured reactive and active powers. Instead of the traditional centralized control structure, this chapter introduces the distributed cooperative control strategy, which only allows each DER agent to communicate with its neighbors. In this way, the nominal points v^{nom} and ω^{nom} used in primary control can be iterated over time in secondary control process.The distributed intelligent cooperative protocol design of voltage and frequency is similar to a pining synchronization problem, while the design of active/reactive power sharing is similar to an average consensus problem. The following text describes detailed distributed cooperative intelligence.

Considering a microgrid consisting of N DER agents, rewrite (4.1) in the form of the droop control law for ith DER agent in (4.2) in the d-q reference coordinate system [191].

$$\begin{cases} \begin{cases} v_i^{\mathrm{od}} = v_i^{\mathrm{nom}} - n_i^Q Q_i \\ v_i^{\mathrm{oq}} = 0 \end{cases} \\ \omega_i = \omega_i^{\mathrm{nom}} - m_i^P P_i \end{cases} \tag{4.2}$$

where v_i^{nom} and ω_i^{nom} are the nominal set-points of the output voltage and frequency on the d-axis, respectively. Active power P_i and reactive power Q_i can be calculated by

$$\begin{cases} P_i = \dfrac{\omega_i^{\mathrm{c}}}{\omega_i^{\mathrm{c}} + s} P_i^{\mathrm{m}} = \dfrac{\omega_i^{\mathrm{c}}}{\omega_i^{\mathrm{c}} + s} \left(v_i^{\mathrm{od}} i_i^{\mathrm{od}} + v_i^{\mathrm{oq}} i_i^{\mathrm{oq}} \right) \\ Q_i = \dfrac{\omega_i^{\mathrm{c}}}{\omega_i^{\mathrm{c}} + s} Q_i^{\mathrm{m}} = \dfrac{\omega_i^{\mathrm{c}}}{\omega_i^{\mathrm{c}} + s} \left(v_i^{\mathrm{od}} i_i^{\mathrm{oq}} - v_i^{\mathrm{oq}} i_i^{\mathrm{od}} \right) \end{cases} \tag{4.3}$$

where ω_i^{c} is the cutoff frequency of the first-order low-pass filter installed in the ith DER's power control loop; v_i^{od} and v_i^{oq} are the direct and quadrature components of output voltage, respectively; i_i^{od} and i_i^{oq} are the direct and quadrature components of operating frequency, respectively; P_i^{m} and Q_i^{m} are the instantaneous active and reactive power components calculated from the measured output voltage and output current [190].

The ith DER agent's control process consists of three loops from the outside to the inside, namely the power, voltage, and current control loops. Based on the nominal set-point ω_i^{nom} and v_i^{nom} of the input, the power control loop generates a reference voltage v_i^{ref}. Similarly, the voltage controller calculates the reference current i_i^{ref} for current control loop from the input reference voltage v_i^{ref}. Finally, the output of the current control loop is converted into pulse signal by the sinusoidal pulse width modulation (SPWM), which is applied to control the inverter.

Based on the reality that the dynamics of the voltage and current control loop are faster than the dynamics of the power control loop [40, 190], time separation operation needs to be introduced in the design of the inverter system. This chapter focuses on the stability of the power control loop, while both the voltage and current control loops use traditional PI controllers.

Differentiating the equations in (4.2) yields

$$\begin{cases} \dot{v}_i^{\text{nom}} = \dot{v}_i^{\text{od}} + n_i^Q \dot{Q}_i \\ \qquad = u_i^v + n_i^Q u_i^Q \\ \dot{\omega}_i^{\text{nom}} = \dot{\omega}_i + m_i^P \dot{P}_i \\ \qquad = u_i^\omega + m_i^P u_i^P \end{cases} \tag{4.4}$$

where u_i^v and u_i^ω are the pinning controllers of voltage and frequency, and u_i^P and u_i^Q are the average consensus controllers of active and reactive power, respectively.

$$\lim_{t \to +\infty} \left| \sum\nolimits_{i=1}^{N} \mu_i v_i^{\text{od}}(t) - v^{\text{DES}} \right| = 0 \tag{4.5}$$

$$\lim_{t \to +\infty} \left| \omega_i(t) - \omega^{\text{DES}} \right| = 0 \tag{4.6}$$

As shown in (4.7), the main control objective of the secondary control level is to design a distributed intelligent cooperative protocol u_i^v, u_i^ω, u_i^P and u_i^Q to regulate the nominal set-points v_i^{nom} and ω_i^{nom} for $i = 1, 2, \cdots, N$ and

$$\lim_{t \to +\infty} \left| \frac{P_i(t)}{P_{i,\max}} - \frac{P_j(t)}{P_{j,\max}} \right| = 0 \tag{4.7}$$

$$\lim_{t \to +\infty} \left| \frac{Q_i(t)}{Q_{i,\max}} - \frac{Q_j(t)}{Q_{j,\max}} \right| = 0 \tag{4.8}$$

for all $i \neq j \in 1, 2, \cdots, N$, where v^{DES} and ω^{DES} are respectively the constant reference voltage and frequency, while $\mu = (\mu_1, \cdots, \mu_N)^T$ is the normalized positive weighting. $P_{i,\max}$ and $Q_{i,\max}$ are the instantaneous maximum capacity of the active and reactive powers, respectively. Equations (4.5)–(4.8) represent the design of distributed droop frequency/voltage control law and consensus active/reactive power control law, respectively. Among them, (4.5) the DER agent's weighted average voltage and DER agent's frequency are asymptotically convergent to a given reference value, and (4.7) the power sharing of active and reactive power is realized.

4.1.1 Voltage and Frequency Restoration

In this section, we design a distributed voltage observer and construct a spinning-based frequency and voltage controller to spin the DER agents' weighted average voltage and frequency to a reference value. Because voltage regulation and reactive power sharing accuracy are difficult to achieve at the same time, a trade-off needs to be made between the two aspects [213]. However, the application of low-voltage microgrids with low rating of DER agents and small electrical distance between units usually lacks static compensation. Therefore, in this case, accurate reactive power sharing between DER agents is crucial, mainly to prevent overload. Therefore, we put the priority of accurate reactive power sharing in the first place, followed by ensuring that the weighted average voltage converges to the reference voltage.

Based on the above description, a distributed voltage observer is designed for each DER agent, assuming that each DER agent can only exchange observations with its neighbors and update its own observations by

$$\hat{v}_i^{\mathrm{od}}(t) = v_i^{\mathrm{od}}(t) + \int_{t_0}^{t} \sum_{j \in N_i^v} a_{ij}^v(r) \cdot \left[\hat{v}_j^{\mathrm{od}}\left(r - \tau_j(r)\right) - \hat{v}_i^{\mathrm{od}}(r - \tau_i(r))\right] \mathrm{d}r \tag{4.9}$$

for $i = 1, 2, \cdots, N$, where $\hat{v}_{j \in N_i^v}^{\mathrm{od}}$ is the ith DER agent's neighbors' observations and $A^v = (a_{ij}^v)_{N \times N}$ is the adjacency matrix. The time varying delays $\tau_k(t) \geq 0$ with $1 \leq k \leq N$ are differentiable functions.

Obviously, the initial observations $\hat{v}_i^{\mathrm{od}}(t_0) = v_i^{\mathrm{od}}(t_0)$. A voltage variation in the ith DER agent immediately results in its own observed response and transmits this effect through the communication link. In this way, the observations of other observers in the communication link network are also affected. Considering the uncertainty in communication links, the digraphs is assumed to be time varying. Thus, we define the switching signal $\sigma(t) : [0, +\infty) \longrightarrow \rho_\Gamma = 1, \cdots, m$. This determines the communication topology G, where $m \in Z^+$ denotes the total number of all possible digraphs in the time varying process.

Now, we deduce the following proposition for the voltage observer (4.9) with time varying delays and variant topologies.

Proposition 4.1 *First, the assumption that G_σ^v is strongly connected to any switching signal $\sigma(t)$ is set. If time varying delays can satisfy both $\tau_k(t) < d$ and $\dot{\tau}_k(t) < d_1 < 1$, where d and d_1 are definite positive constants for all $1 \leq k \leq N$, then voltage observer* (4.9) *can asymptotically converge DERs' voltage observations to a weighted average value of all DERs' actual voltage.*

$$\lim_{t \to +\infty} \hat{v}_i^{\mathrm{od}} = \textstyle\sum_{i=1}^{N} \mu_i \hat{v}_i^{\mathrm{od}}(t), i = 1, \cdots, N \tag{4.10}$$

where $\mu = (\mu_1, \cdots, \mu_N)^T$ is the positive left eigenvector which is normalized corresponding to the zero eigenvalue of Laplacian matrix L_σ^v of G_σ^v.

Proof In the case of fixed switching signal $\sigma(t)$, by differentiating (4.9) and (4.11) can be obtained

$$\dot{\hat{v}}_i^{\mathrm{od}}(t) = \dot{v}_i^{\mathrm{od}}(t) + \sum_{j \in N_i^v(s)} a_{ij}^v \left[\hat{v}_j^{\mathrm{od}}\left(t - \tau_j(t)\right) - \hat{v}_i^{\mathrm{od}}\left(t - \tau_i(t)\right)\right] \tag{4.11}$$

where $\hat{v}^{od} = (\hat{v}_1^{od}, \cdots \hat{v}_N^{od})^T$, $v^{od} = (v_1^{od}, \cdots, v_N^{od})^T$, $L_\sigma^v = (a_{ij}^v)_{N\times N}$ and $\tau(t) = (\tau_1(t), \cdots, \tau_N(t))^T$. Therefore, the global observer dynamics can be modeled as

$$\dot{\hat{v}}^{od}(t) = \dot{v}^{od}(t) - L_\sigma^v \hat{v}^{od}(t - \tau(t)) \tag{4.12}$$

Convert (4.12) to the frequency domain form

$$\hat{U}^{od}(s) = s\left[sI_N + L_\sigma^v\left(e^{-s\tau(\vartheta)}\right)/(1 - \dot{\tau}(\vartheta))\right]^{-1} U^{od}(s) \tag{4.13}$$

where ϑ is some fixed positive constant, $\hat{U}^{od}$ and U^{od} are Laplace performs of $\dot{v}^{od}$ and v^{od}, respectively.

According to Nyquist stability criterion, we derive the stability condition of transfer function $s\left[sI_N + L_\sigma^v\left(e^{-s\tau(\vartheta)}\right)/(1 - \dot{\tau}(\vartheta))\right]^{-1}$: $d < \frac{(1-d_1)\pi}{2\max_{i=2,\cdots,N}|\lambda_i(L_\sigma^V)|}$ Next, we analyze the linear dynamics of the above transfer function $\dot{x}(t) = -L_\sigma^v x(t-\tau(t))$. Because digraphs G_σ^v are strongly connected, there exists a positive left eigenvector corresponding to zero eigenvalues of L_σ^v. Therefore, the value of $\sum_{i=1}^N \mu_i x_i(t)$ does not vary during the iteration. According to the final value theorem, we can derive the (4.10). The proof is completed.

Then, in order to make the d-axis voltage u_i^v and frequency u_i^ω track the reference values v^{DES} and ω^{DES} respectively, we design two pining controllers. According to the previous assumption, each DER agent in the communication network can only communicate with its own neighbors, we can obtain (4.14)

$$\begin{cases} u_i^v = \sum_{j\in N_i^v(t)} a_{ij}^v(t)[\hat{v}_j^{od}(t-\tau_j(t)) - \hat{v}_i^{od}(t-\tau_i(t))] \\ \quad + a_{i0}^v(t)[v^{DES} - \hat{v}_i^{od}(t-\tau_i(t))] \\ u_i^\omega = \sum_{j\in N_i^\omega(t)} a_{ij}^\omega(t)[\omega_j(t-\tau_{j+N}(t)) - \omega_i(t-\tau_{i+N}(t))] \\ \quad + a_{i0}^\omega(t)[\omega^{DES} - \omega_i(t-\tau_{i+N}(t))] \end{cases} \tag{4.14}$$

for $i = 1, 2, \cdots, N$, where the time varying delays $\tau_k(t) \geq 0, 1 \leq k \leq 2N$ are continuously differentiable functions. At the same time, assume that the elements of the adjacency matrices $A^v(t) = (a_{ij}^v(t))_{N\times N}$ and $A^\omega(t) = (a_{ij}^\omega(t))_{N\times N}$ are time varying. $G^v(t)$ and $G^\omega(t)$ are time varying digraphs corresponding to the adjacency matrix $A^v(t)$ and $A^\omega(t)$, respectively. We assume that the delay function and adjacency matrix for voltage and frequency control are different because they come from different dynamic systems.

Theorem 4.1 *Assume that all directed graphs G_σ^v and G_σ^ω are strongly connected, and ensure that at least one element in the voltage and frequency leader adjacency matrix satisfies $a_{i0}^v > 0$ and $a_{i0}^\omega > 0$, which means at least one follower connects to the leader. If time varying function $\tau_k(t)$ satisfies* (4.15), *the inequality* (4.16) *holds for some system matrices Ξ, H_σ and L_σ^v. In this way, the pinning-based controller*

(4.14) *can pin all DERs' weighted average voltage* $\sum_{i=1}^{N}\mu_i v_i^{\mathrm{od}}(t)$ *and frequency* ω_i *to reference value* v^{DES} *and* ω^{DES}.

$$\begin{aligned} \tau_k(t) &< d \\ \dot{\tau}_k(t) &< d_1 < 1 \end{aligned} \tag{4.15}$$

where d *and* d_1 *are positive constants, which can be represented by*

$$0 < d < \min\left\{\frac{(1-d_1)\pi}{2\max_{i=2,\cdots,N}\left|\lambda_i\left(L_\sigma^v\right)\right|} \right. \\ \left. \frac{(1-d_1)\lambda_{\min}\left(\Xi H_\sigma + H_\sigma^T \Xi\right)}{\lambda_{\max}\left(H_\sigma^T H_\sigma\right)+(1-d_1)\lambda_{\max}\left(\Xi H_\sigma H_\sigma^T \Xi\right)}\right\} \tag{4.16}$$

Proof For the convenience of analysis, parameters $B_\sigma^v = \mathrm{diag}\left\{a_{10}^v, \cdots, a_{N0}^v\right\}$, $e(t) = \xi(t) - \bar{\xi} = (e_1(t), \cdots, e_N(t), e_{N+1}(t), \cdots, e_{2N}(t)) \in R^{2N}$ are set in advance, where $\xi = (v_1^{\mathrm{od}}, \cdots, v_N^{\mathrm{od}}, \omega_1, \cdots, \omega_N)^T \in R^{2N}$ and $\bar{\xi} = ((v^{\mathrm{ref}}1_N)^T, (\omega^{\mathrm{ref}}1_N)^T)^T$.

By referring to Eqs. 4.4 and 4.14 and fact $\dot{v}^{\mathrm{DES}} = \dot{\omega}^{\mathrm{DES}} = 0$, we can derive:

$$\dot{e}_l(t) = -\sum_{k=1}^{2N} h_{lk}^\sigma e_k\left(t-\tau_k(t)\right), l,k = 1,\cdots,2N \tag{4.17}$$

where $H_\sigma = \left(h_{lk}^\sigma\right)_{2N\times 2N}$ =diag $\left\{H_\sigma^v, H_\sigma^\omega\right\}$ with $H_\sigma^v = L_\sigma^v + B_\sigma^v$ and $H_\sigma^\omega = L_\sigma^\omega + B_\sigma^\omega$. These two are related to the Laplace matrices L_σ^v and L_σ^ω of digraph G_σ^v and G_σ^ω. Define the Lyapunov-Krasovskii function for the system shown in (4.17):

$$\begin{aligned} V_1(t) =& \sum_{k=1}^{2N} q_l e_l^2(t) \\ &+ \int_{-d}^{0}\int_{t+\theta}^{t}\sum_{l=1}^{2N}\dot{e}_l^2(r)\mathrm{d}r\mathrm{d}\theta \\ &+\gamma\int_{t-\tau_k(t)}^{t}\sum_{l=1}^{2N}\left(\sum_{k=1}^{2N}h_{lk}^\sigma e_k(r)\right)^2\mathrm{d}r \end{aligned} \tag{4.18}$$

where $(q_1,\cdots,q_N)^T$ and $(q_{N+1},\cdots,q_{2N})^T$ are the positive left eigenvectors corresponding to the zero eigenvalue of the Laplacian matrices L_σ^v and L_σ^ω. We choose the $\sum_{l=1}^{N} q_l = 1$, $\sum_{l=N+1}^{2N} q_l = 1$, and $q_l > 0$ to guarantee the strong connectivity of the digraph G_σ^v and G_σ^ω for all $l = 1, \cdots, 2N$.

Taking the derivative of the $V_1(t)$ yields:

$$
\begin{aligned}
\dot{V}_1(t) = & -2\sum_{l=1}^{2N}\sum_{k=1}^{2N} q_l e_l(t) h_{lk}^{\sigma} e_k\left(t-\tau_k(t)\right) - \int_{t-d}^{t}\sum_{l=1}^{2N}\dot{e}_l^2(r)\mathrm{d}r \\
& + \gamma\sum_{l=1}^{2N}\left(\sum_{k=1}^{2N} h_{lk}^{\sigma} e_k(t)\right)^2 \\
& - \gamma\left(1-\dot{\tau}_k(t)\right)\sum_{l=1}^{2N}\left(\sum_{k=1}^{2N} h_{lk}^{\sigma} e_k\left(t-\tau_k(t)\right)\right)^2 + d\sum_{l=1}^{2N}\dot{e}_l^2(t),
\end{aligned}
\tag{4.19}
$$

A suitable positive definite matrix A is set to ensure that $2x^T y \leq x^T A^{-1} x + y^T A y$ holds, and according to Newton-Leibniz formula, it can be obtained:

$$
\begin{aligned}
& -2\sum_{l=1}^{2N}\sum_{k=1}^{2N} q_l e_l(t) h_{lk}^{\sigma} e_k\left(t-\tau_k(t)\right) \\
= & -2\sum_{l=1}^{2N}\sum_{k=1}^{2N} q_l e_l(t) h_{lk}^{\sigma} e_k(t) \\
& + \int_{t-\tau_k(t)}^{t} 2\sum_{l=1}^{2N}\sum_{k=1}^{2N} q_l e_l(t) h_{lk}^{\sigma} \dot{e}_k(r)\mathrm{d}r \\
= & -e^T(t)\left(\Xi H_\sigma + H_\sigma^T \Xi\right) e(t) \\
& + \int_{t-\tau_k(t)}^{t} 2e^T(t)\Xi H_\sigma \dot{e}(r)\mathrm{d}r \\
\leq & -e^T(t)\left(\Xi H_\sigma + H_\sigma^T \Xi\right) e(t) \\
& + \tau_k(t) e^T(t)\left(\Xi H_\sigma H_\sigma^T \Xi\right) e^T(t) \\
& + \int_{t-\tau_k(t)}^{t} \dot{e}^T(r)\dot{e}(r)\mathrm{d}r \\
\leq & -e^T(t)\left(\Xi H_\sigma + H_\sigma^T \Xi\right) e(t) \\
& + d e^T(t)\left(\Xi H_\sigma H_\sigma^T \Xi\right) e^T(t) \\
& + \int_{t-d}^{t}\sum_{l=1}^{2N}\dot{e}_k^2(r)\mathrm{d}r,
\end{aligned}
\tag{4.20}
$$

where $\Xi = \mathrm{diag}\{q_1, \cdots, q_{2N}\}$.

Further derivation leads to:

$$\begin{aligned}
\dot{V}_1(t) \leq & -e^T(t)\left(\Xi H_\sigma + H_\sigma^T \Xi\right) e(t) \\
& + d e^T(t)\left(\Xi H_\sigma H_\sigma^T \Xi\right) e^T(t) \\
& + \gamma \sum_{l=1}^{2N} \left[\sum_{k=1}^{2N} h_{lk}^\sigma e_k(t)\right]^2 \\
& + d \sum_{l=1}^{2N} \left[\sum_{k=1}^{2N} h_{lk}^\sigma e_k\left(t - \tau_k(t)\right)\right]^2 \\
& - \gamma\left(1 - d_1\right) \sum_{l=1}^{2N} \left[\sum_{k=1}^{2N} h_{lk}^\sigma e_k\left(t - \tau_k(t)\right)\right]^2 \\
= & -e^T(t)\left(\Xi H_\sigma + H_\sigma^T \Xi\right) e(t) \\
& + d e^T(t)\left(\Xi H_\sigma H_\sigma^T \Xi\right) e^T(t) \\
& + d Y^T(t)\left(H_\sigma^T H_\sigma\right) Y(t) \\
& + \gamma e^T(t)\left(H_\sigma^T H_\sigma\right) e(t) \\
& - \gamma\left(1 - d_1\right) Y^T(t)\left(H_\sigma^T H_\sigma\right) Y(t)
\end{aligned} \tag{4.21}$$

where $Y(t) = \left(y_1(t), \cdots, y_{2N}(t)\right)^T$ with $y_k(t) = e_k\left(t - \tau_k(t)\right)$ for $1, 2, \cdots, 2N$.

Then, it can be obtained that

$$\begin{aligned}
\dot{V}_1(t) \leq & -\Big[\lambda_{\min}\left(\Xi H_\sigma + H_\sigma^T \Xi\right) - d\lambda_{\max}\left(\Xi H_\sigma H_\sigma^T \Xi\right) \\
& -\gamma \lambda_{\max}\left(H_\sigma^T H_\sigma\right)\Big] e^T(t) e(t) \\
& -\left[\left(\gamma\left(1 - d_1\right) - d\right) \lambda_{\min}\left(H_\sigma^T H_\sigma\right)\right] Y^T(t) Y(t)
\end{aligned} \tag{4.22}$$

Therefore, conditions for the establishment of $\dot{V}_1(t) < 0$ are as follows:

$$\begin{cases}
\lambda_{\min}\left(\Xi H_\sigma + H_\sigma^T \Xi\right) - d\lambda_{\max}\left(\Xi H_\sigma H_\sigma^T \Xi\right) - \gamma \lambda_{\max}\left(H_\sigma^T H_\sigma\right) > 0 \\
\left[\gamma\left(1 - d_1\right) - d\right] \lambda_{\min}\left(H_\sigma^T H_\sigma\right) > 0
\end{cases} \tag{4.23}$$

which leads to:

$$\frac{d}{1-d_1} < \gamma < \frac{\lambda_{\min}\left(\Xi H_\sigma + H_\sigma^T \Xi\right) - d\lambda_{\max}\left(\Xi H_\sigma H_\sigma^T \Xi\right)}{\lambda_{\max}\left(H_\sigma^T H_\sigma\right)} \tag{4.24}$$

Then, it can be deduced that:

$$d < \frac{(1-d_1)\,\lambda_{\min}\left(\Xi H_\sigma + H_\sigma^T \Xi\right)}{\lambda_{\max}\left(H_\sigma^T H_\sigma\right) + (1-d_1)\,\lambda_{\max}\left(\Xi H_\sigma H_\sigma^T \Xi\right)} \tag{4.25}$$

Given that $d_1 < 1$, we can derive the conclusion that $\lim_{t\to+\infty}\left|\hat{v}_i^{\text{od}}(t) - v^{\text{DES}}\right| = 0$. Then, $\lim_{t\to+\infty}\left|\sum_{i=1}^{N}\mu_i\hat{v}_i^{\text{od}}(t) - v^{\text{DES}}\right| = 0$ can be obtained by combining the conclusion in Proposition 4.1. This completes the proof.

Remark 4.1 Although the proposed pinning-based distributed controller takes into account the existence of virtual leaders, leader failures may occur due to the uncertainty of the communication network. In the case of leader failure, we design a local emergency strategy for each DER agent of the communication network species, in which the reference value of each DER agent can be approximated as (4.26).

$$\begin{cases} \bar{v}_i^{\text{DES}} = v_i^{\text{od}} + \dfrac{\sum_{j\in N_i^v} a_{ij}^v \left(v_j^{\text{od}} - v_i^{\text{od}}\right)}{\sum_{j\in N_i^v} a_{ij}^v} \\[2ex] \bar{\omega}_i^{\text{DES}} = \omega_i + \dfrac{\sum_{j\in N_i^\omega} a_{ij}^\omega \left(\omega_j - \omega_i\right)}{\sum_{j\in N_i^\omega} a_{ij}^\omega} \end{cases} \tag{4.26}$$

Each DER agent can be regarded as an agent, when it is disconnected from the leader, it will regulate its own state according to the state of its neighbors, reflecting a kind of self-organization idea. In some extreme cases, if the value of a variable exceeds a threshold, the system will enable a self-protection policy for each DER agent such as black start [216, 217].

Remark 4.2 Since the physical properties of the electromagnetic system do not change with the coordinate system, the proposed distributed control strategy can be extended to the stationary coordinate frame without changing the analysis results. In this way, the analysis process can be simplified by simply converting the coordinates of the synchronous rotating coordinate frame to the stationary coordinate frame [190].

Remark 4.3 In the stability analysis of non-uniform delay communication networks, the influence of delay is not considered or the delay is sufficiently small. In order to obtain the upper bound of the non-uniform time varying delay, an Lyapunov-Krasovskii function containing the delay dependent parameters is constructed in this chapter. The difficulty of constructing Lyapunov-Krasovskii function increases mainly because of the asymmetry of adjacency matrix, nonlinearity of control system and non-uniform delay in the existing directed dynamical network topology. This chapter solves the above problems by using the special matrix theory, inequality theory, and theory of delay differential equations.

4.1.2 Active and Reactive Powers Sharing

As shown in (4.27), the proportional utilization of ith DER agent is obtained.

$$\frac{P_i}{P_{i,\max}} = \alpha_i \tag{4.27}$$

$$\frac{Q_i}{Q_{i,\max}} = \beta_i \tag{4.28}$$

where α_i and β_i are the ratio of active and reactive power output of the ith DER, whose instantaneous maximum capacity $P_{i,\max}$ and $Q_{i,\max}$. α_i and β_i represent the utilization percentage of each DER, and the utilization percentage as a control objective needs to be regulated to the same value at steady state, i.e. $\alpha_i = \alpha_j = \alpha^\star$ and $\beta_i = \beta_j = \beta^\star$. Therefore, the distributed network can set different utilization objective so that the usage of each DER converges to the reference value. For example, if $\alpha_i^\star$ is the desired reference value of common ratio, the controller introduces a gain $k_i = \frac{\alpha^\star}{\alpha_i^\star}$ to guide transformed utilization ratio $k_i\alpha_i$ to converge to $\alpha^\star$ [218]. In this way, any specific utilization profile can be translated into a power utilization profile in (4.7). However, in the absence of a virtual leader, if the reference ratio $\alpha^\star$ and $\beta^\star$ are unknown, the average consensus problem needs to be introduced for active/reactive power sharing.

Based on the previous analysis, we design a distributed intelligent cooperative protocol for active /reactive power sharing with two average consensus controllers u_i^P and u_i^Q, which allocate active/reactive power based on the output ratio. Similar to the frequency and voltage controller, the average consensus controller u_i^P and u_i^Q only receives information about each DER agent itself and its neighbors as shown in (4.29).

$$\begin{cases} u_i^P = P_{i,\max} \sum\limits_{j\in N_i^P(t)} a_{ij}^P(t) \left[\dfrac{P_j\left(t - T_j(t)\right)}{P_{j,\max}} - \dfrac{P_i\left(t - T_i(t)\right)}{P_{i,\max}} \right] \\ u_i^Q = Q_{i,\max} \sum\limits_{j\in N_i^Q(t)} a_{ij}^Q(t) \left[\dfrac{Q_j\left(t - T_{j+N}(t)\right)}{Q_{j,\max}} - \dfrac{Q_i\left(t - T_{i+N}(t)\right)}{Q_{i,\max}} \right] \end{cases} \tag{4.29}$$

for $i = 1, \cdots, N$ and continuously differentiable time varying delay functions $T_k(t), 1 \le k \le 2N$.

Theorem 4.2 *It is assumed that the digraphs G_σ^P and G_σ^Q are strongly connected under any switching signal $\sigma(t)$. If the time varying delays satisfy* (4.30)*, inequality* (4.31) *holds for any system matrices ψ and L_σ. In this way, the distributed controller can achieve precise power sharing of active power P_i and reactive power Q_i of the ith DER agent.*

$$
\begin{aligned}
&T_k(t) < \hat{d},\ \dot{T}_k(t) < \hat{d}_1 < 1 \\
&\dot{d} > 0,\ \dot{d}_1 > 0,\ 1 < k < 2N
\end{aligned} \tag{4.30}
$$

$$
\hat{d} < \frac{\left(1-\hat{d}_1\right)\lambda_3\left(\psi L_\sigma + L_\sigma^T \psi\right)}{\lambda_{\max}\left(L_\sigma^T L_\sigma\right) + \left(1-\hat{d}_1\right)\lambda_{\max}\left(\psi L_\sigma L_\sigma^T \psi\right)} \tag{4.31}
$$

Proof Set the aforementioned parameters

$$
\eta = (P_1, \cdots, P_N, Q_1, \cdots, Q_N)^T \in R^{2N} \tag{4.32}
$$

$$
\bar{\eta} = \left(\left(\sum_{j=1}^{N} p_j \frac{P_j(0)}{P_{j,\max}} 1_N\right)^T, \left(\sum_{j=1}^{N} p_{j+N} \frac{Q_j(0)}{Q_{j,\max}} 1_N\right)^T\right)^T \in R^{2N} \tag{4.33}
$$

$$
\delta(t) = \eta_t - \bar{\eta}(t) = (\delta_1(t), \cdots, \delta_N(t))^T \in R^N \tag{4.34}
$$

where $(p_1, \cdots, p_N)^T \in R^N$ and $(p_{N+1}, \cdots, p_{2N})^T \in R^N$ are the left positive eigenvectors corresponding to the zero eigenvalues of the Laplacian matrices L_σ^P and L_σ^Q.

To ensure the strong connectivity of the digraph G_σ^P and G_σ^Q, we set the parameters $\sum_{l=1}^{N} p_l = 1$, $\sum_{l=N+1}^{2N} p_l = 1$ and $p_l > 0$ for all $l = 1, 2, \cdots, 2N$. Combining formulas (4.4) and (4.29) can be derived:

$$
\dot{\delta}_l(t) = -\sum_{k=1}^{2N} \ell_{lk}^\sigma \delta_k\left(t - T_k(t)\right) l, k = 1, \ldots, 2N \tag{4.35}
$$

where $L_\sigma = (\ell_{lk}^\sigma)_{2N\times 2N}$ =diag $\left\{L_\sigma^P, L_\sigma^Q\right\}$ with Laplacian matrices L_σ^P(of digraph G_σ^P) and L_σ^Q(of digraph G_σ^Q).

Define the Lyapunov-Krasovskii function of system (4.35):

$$
\begin{aligned}
V_2(t) =& \sum_{k=1}^{2N} p_l \delta_l^2(t) \\
&+ \int_{-\hat{d}}^{0} \int_{t+\theta}^{t} \sum_{l=1}^{2N} \dot{\delta}_l^2(r) \mathrm{d}r \, \mathrm{d}\theta \\
&+ \gamma \int_{t-T_k(t)}^{t} \sum_{l=1}^{2N} \left[\sum_{k=1}^{2N} \ell_{lk}^{\sigma} \delta_k(r) \right]^2 \mathrm{d}r
\end{aligned}
\tag{4.36}
$$

Similar to the proof of Theorem 3.1, taking the derivative of $V_2(t)$ yields:

$$
\begin{aligned}
\dot{V}_2(t) \leq & - \delta^T(t) \left(\psi L_\sigma + L_\sigma^T \psi \right) \delta(t) \\
& + \gamma \delta^T(t) \left(L_\sigma^T L_\sigma \right) \delta(t) \\
& + \hat{d} \delta^T(t) \left(\psi L_\sigma L_\sigma^T \psi \right) \delta^T(t) \\
& - \gamma \left[\left(1 - \hat{d}_1 \right) - \hat{d} \right] Z^T(t) \left(L_\sigma^T L_\sigma \right) Z(t)
\end{aligned}
\tag{4.37}
$$

where $Z(t) = (z_1(t), \cdots, z_{2N}(t))^T$ with $z_k(t) = \delta_k(t - T_k(t))$ for $k = 1, \cdots, 2N$ and $\psi = \mathrm{diag}\{p_1, \cdots, p_{2N}\}$.

Set $\Omega = ((\phi(t)1_N)^T, (\varphi(t)1_N)^T)^T$, where $\phi(t)1_N$ and $\varphi(t)1_N$ represent the first N and last N elements of $\delta(t)$ to the subspace spanned by 1_N. Then, we decompose $\delta(t) = \Omega \oplus \bar{\delta}(t)$ with $1_N^T \delta(t) = 0$ using spatial orthogonal decomposition method. Since $\psi L_\sigma + L_\sigma^T \psi$, $\psi L_\sigma L_\sigma^T \psi$, and $L_\sigma^T L_\sigma$ are symmetric matrices and the row sum is 0, the eigenvalues of $\psi L_\sigma + L_\sigma^T \psi$ and $L_\sigma^T L_\sigma$ can be set to $0 = \lambda_1(\cdot) < \lambda_2 < \lambda_3(\cdot) < \cdots < \lambda_N(\cdot)$. Thus,

$$
\begin{aligned}
\dot{V}_2(t) \leq & - \lambda_3 \left(\psi L_\sigma + L_\sigma^T \psi \right) \bar{\delta}^T(t) \bar{\delta}(t) \\
& + \left[\gamma \lambda_{\max} \left(L_\sigma^T L_\sigma \right) + \hat{d} \lambda_{\max} \left(\psi L_\sigma L_\sigma^T \psi \right) \right] \bar{\delta}^T(t) \bar{\delta}(t) \\
& - \gamma \left[\left(1 - \hat{d}_1 \right) - \hat{d} \right] \lambda_3 \left(L_\sigma^T L_\sigma \right) \bar{Z}^T(t) \bar{Z}(t),
\end{aligned}
\tag{4.38}
$$

where $\bar{Z}(t) = (\bar{z}_1(t), \cdots, \bar{z}_{2N}(t))^T$ with $\bar{z}_k(t) = \bar{\delta}_k(t - T_k(t))$ for all k. Setting $\dot{V}_2(t) < 0$ yields:

$$
\begin{cases}
\lambda_3 \left(\psi L_\sigma + L_\sigma^T \psi \right) - \hat{d} \lambda_{\max} \left(\psi L_\sigma L_\sigma^T \psi \right) - \gamma \lambda_{\max} \left(L_\sigma^T L_\sigma \right) > 0 \\
\left[\gamma \left(1 - \hat{d}_1 \right) - \hat{d} \right] \lambda_3 \left(L_\sigma^T L_\sigma \right) > 0
\end{cases}
\tag{4.39}
$$

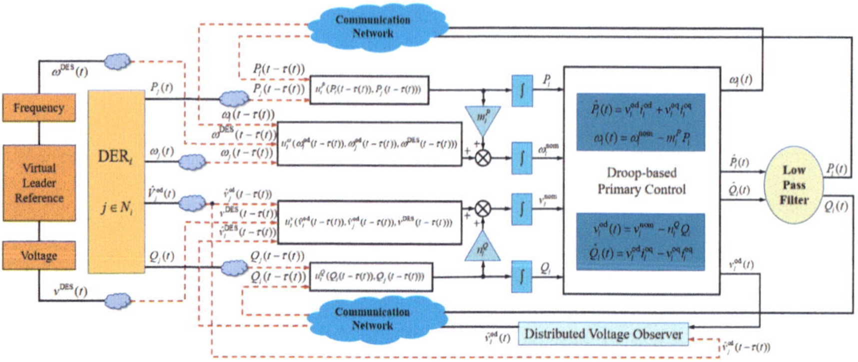

Fig. 4.1 The diagram of droop-based distributed intelligent cooperative protocol

Finally, it is deduced that for given d_1:

$$\hat{d} < \frac{\left(1 - \hat{d}_1\right) \lambda_3 \left(\psi L_\sigma + L_\sigma^T \psi\right)}{\lambda_{\max}\left(L_\sigma^T L_\sigma\right) + \left(1 - \hat{d}_1\right) \lambda_{\max}\left(\psi L_\sigma L_\sigma^T \psi\right)} \tag{4.40}$$

This completes the proof.

As shown in the Fig. 4.1, we design a block diagram of droop-based distributed cooperative control for voltage/frequency planning and active/reactive power sharing. In the primary control stage, the droop-based technique is actually implemented in the internal control loop of each DER agent, with the nominal set point generated by (4.41). In the secondary control stages shown in (4.14) and (4.29), the control output v_i^{od}, ω_i, P_i and Q_i of each DER agent are transmitted to its neighbors through a communication network that contains non-uniform time varying delays $\tau_k(t)$ and $T_k(t)$. The external given reference value is sent to a specific DER agent that is connected to the virtual leader. Through distributed communication networks, each DER agent transmits its own information to its neighbors, and eventually the active and reactive power ratio α_i and β_i of all DER agents will achieve consensus to ensure accurate power sharing.

$$\begin{cases} v_i^{\text{nom}} = \displaystyle\int (u_i^v + n_i^Q u_i^Q) dt \\ \omega^{\text{nom}} = \displaystyle\int (u_i^\omega + m_i^P u_i^P) dt \end{cases} \tag{4.41}$$

Remark 4.4 The control strategy proposed in this chapter is different from the existing research. The problem of voltage regulation is mainly considered in [191], while the problem of reactive power sharing is not studied in [107], and the communication delay in the design of frequency/voltage restoration and control of active power sharing is also not considered. Although the influence of communication delay on reactive power sharing is reflected in the experimental results in [40], the dynamics and stability of the control system are not analyzed in detailed.

Remark 4.5 This section uses a distributed communication network, allowing each DER agent to obtain key information only from its neighbors. This is different from the requirement for DER agents to obtain global information in [40], which significantly reduces the communication cost. At the same time, the distributed voltage observer designed in this chapter not only ensures precise reactive power sharing, but also ensures that the weighted average of DER agents can asymptotically converge to the reference value. This solves the inherent contradiction between the two, which is not realized in [40].

Remark 4.6 By applying the proposed distributed intelligent cooperative protocol for voltage regulation (4.9) and distributed voltage observer (4.11), the weighted average of all DER agents converges asymptotically to the reference value. The weighting factor $\mu = (\mu_1, \cdots, \mu_N)^T$ is the normalized positive left eigenvector corresponding to the zero eigenvalue of the Laplace matrix L^v (related to the G^v). In practice, we choose G^v balanced and strongly connected digraph, and by adjusting the value of μ, the average of all DER agents converges to a desire reference value. In this sense, we give the following corollary.

Remark 4.7 The proposed control framework effectively decouples the voltage regulation and reactive power sharing objectives, which are often conflicting in conventional droop based microgrid systems. By introducing a distributed average voltage observer, each DG can locally estimate the global average value of voltage amplitudes without requiring direct measurement from all DERs. This design not only mitigates the impact of line impedance mismatches but also enhances the systems adaptability to varying operating conditions. Such a feature is particularly beneficial in low-voltage microgrids, where accurate reactive power sharing is critical to prevent DG overloading.

Remark 4.8 It is worth emphasizing that the communication topology considered in this chapter is both directed and switching, which better reflects the practical constraints of microgrids. Unlike undirected or fixed topologies, the proposed approach accommodates intermittent and asymmetric information exchange, thereby supporting plug-and-play operations and resilience to link failures. The use of the pinning based control strategy further reduces the dependency on continuous global communication, which will be a significant advantage in scalable systems.

Corollary 4.1 *Assuming that the conditions of Theorems 3.1–3.2 are satisfied, the digraph G^v_σ is balanced for any switching signal $\sigma(t)$. Distributed cooperative protocols* (4.14) *and* (4.29) *and distributed voltage observers* (4.9) *ensure that the voltages and frequencies of all DER agents converge progressively to the desired reference values while maintaining active and reactive power sharing.*

Proof Because the digraph G^v_σ needs to be balanced and strongly connected, regardless of the switching signal $\sigma(t)$. According to the property of balanced graph, the left eigenvector $1 = (1, \cdots, 1)^T$ corresponding to the zero eigenvalue of Laplace matrix can be obtained. After $1 = (1, \cdots, 1)^T$ normalization, $\mu = (1/N, \cdots 1/N)^T$ is obtained, and the proof is completed by combining Proposition 4.1.

4.2 Simulation Results and Discussion

In this chapter, simulation will be expressed to verify the effectiveness of the proposed distributed control algorithms. First, the 5 DER agents microgrid shown in Fig. 4.2 is modeled in MATLAB/SimPowerSystems. The electrical parameters of DER agent, load and line and the control policy parameters of DER agent are shown in Table 4.1. The communication network can be simplified into the topology of $\hat{G}_1$ in Fig. 4.3. The associated Laplacian matrices of $\hat{G}_1$ are

$$L_1^v = 0.01 \times \begin{pmatrix} 0.5 & -0.25 & 0 & 0 & -0.25 \\ -0.25 & 0.5 & -0.25 & 0 & 0 \\ 0 & -0.2 & 0.4 & -0.2 & 0 \\ 0 & 0 & -0.15 & 0.15 & 0 \\ 0 & 0 & 0 & -0.25 & 0.25 \end{pmatrix} \tag{4.42}$$

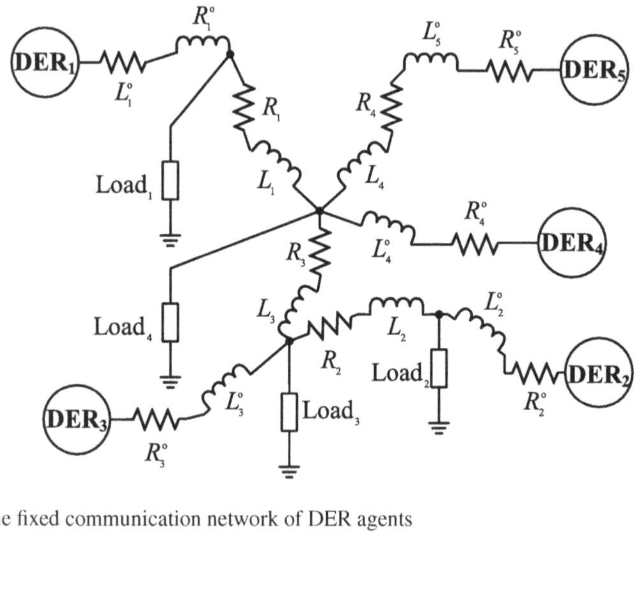

Fig. 4.2 The fixed communication network of DER agents

Table 4.1 The parameters values for the microgrid system

DERs	Parameter	$DER_{1\&4}$	Parameter	$DER_{2\&3\&5}$
	V^{DC}	800 V	V^{DC} V	680 V
	m^{P}	1.06×10^{-4}	m^{P}	0.708×10^{-4}
	n^{Q}	5.02×10^{-3}	n^{Q}	5.26×10^{-3}
	L^{f}	0.15 mH	L^{f}	0.15 mH
	C^{f}	50 μF	C^{f}	50 μF
	R^{f}	0.05 Ω	R^{f}	0.05 Ω
	L^{o}	0.21 mH	L^{o}	0.175 mH
	R^{o}	0.05 Ω	R^{o}	0.06 Ω
	P^{max}	20 kW	P^{max}	30 kW
	Q^{max}	13.2 kVar	Q^{max}	22 kVar
Loads	$Load_1$	$Load_2$	$Load_3$	$Load_4$
	P_1 19.4 kW	P_2 27.8 kW	P_3 22.4 kW	P_4 21.3 kW
	Q_1 13 kVar	Q_2 18.4 kVar	Q_3 15.5 kVar	Q_4 13.5 kVar
Lines	$Line_1$	$Line_2$	$Line_3$	$Line_4$
	R_1 0.18 Ω	R_2 0.34 Ω	R_3 0.2 Ω	R_1 0.3 Ω
	L_1 2.9 μH	L_2 5.6 μH	L_3 3.2 μH	L_4 4.9 μH

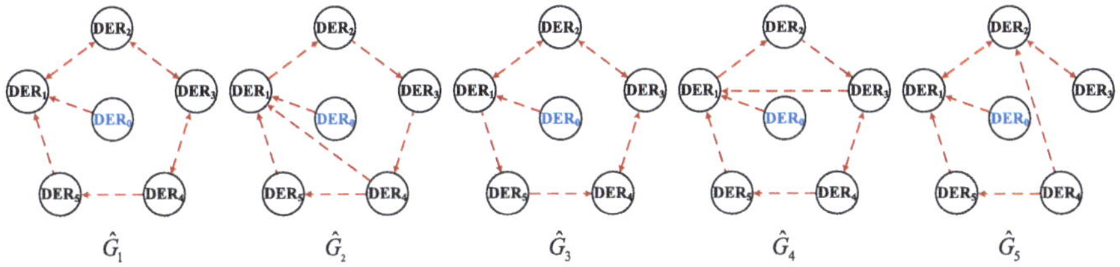

Fig. 4.3 The switching communication digraph topology

$$L_1^{\omega} = 0.01 \times \begin{pmatrix} 1 & -0.5 & 0 & 0 & -0.5 \\ -0.5 & 1 & -0.5 & 0 & 0 \\ 0 & -0.35 & 0.7 & -0.35 & 0 \\ 0 & 0 & -0.45 & 0.45 & 0 \\ 0 & 0 & 0 & -0.5 & 0.5 \end{pmatrix} \quad (4.43)$$

$$L_1^{P} = \begin{pmatrix} 0.064 & -0.032 & 0 & 0 & -0.032 \\ -0.04 & 0.08 & -0.04 & 0 & 0 \\ 0 & -0.012 & 0.024 & -0.012 & 0 \\ 0 & 0 & -0.024 & 0.024 & 0 \\ 0 & 0 & 0 & -0.04 & 0.04 \end{pmatrix} \quad (4.44)$$

$$L_1^Q = \begin{pmatrix} 0.16 & -0.08 & 0 & 0 & -0.08 \\ -0.032 & 0.064 & -0.032 & 0 & 0 \\ 0 & -0.056 & 0.112 & -0.056 & 0 \\ 0 & 0 & -0.064 & 0.064 & 0 \\ 0 & 0 & 0 & -0.08 & 0.08 \end{pmatrix} \tag{4.45}$$

And the value of the corresponding leader adjacency matrix are $B_1^v = diag\{0.0125, 0, 0, 0, 0\}$ and $B_1^\omega = diag\{0.0225, 0, 0, 0, 0\}$. As shown in digraph $\hat{G}_1$ of the Fig. 4.3, only DER_1 agent can obtain reference values from the virtual leader, and other DER agents can only obtain reference values through distributed communication with neighbors. Digraph $\hat{G}_2 - \hat{G}_5$ also has the same communication mode.

The parameters are set for the following simulation verification

$$\tau_k(t) = \frac{1 + 0.1\sin(t + k)}{16}, \text{ for } k = 1, 2\cdots, 10 \tag{4.46}$$

$$T_k(t) = \frac{1 + 0.2\sin(2t + 2k)}{16}, \text{ for } k = 1, 2\cdots, 10 \tag{4.47}$$

then, it can be obtained that

$$\tau_k(t) < 0.0688 < d \tag{4.48}$$

$$\dot{\tau}_k(t) < 0.0063 < d_1 < 1 \tag{4.49}$$

and we have

$$T_k(t) < 0.075 < \hat{d} \tag{4.50}$$

$$\dot{T}_k(t) < 0.0125 < \hat{d}_1 < 1 \tag{4.51}$$

where $d, d_1, \hat{d}$ and $\hat{d}_1$ are positive constants.

Sets the parameters of the fixed digraph $\hat{G}_1$ as

$$\Xi = \begin{bmatrix} 0.0694 & 0 & 0 & 0 & 0 & 0 & 0 & 0 & 0 & 0 \\ 0 & 0.1387 & 0 & 0 & 0 & 0 & 0 & 0 & 0 & 0 \\ 0 & 0 & 0.2601 & 0 & 0 & 0 & 0 & 0 & 0 & 0 \\ 0 & 0 & 0 & 0.4624 & 0 & 0 & 0 & 0 & 0 & 0 \\ 0 & 0 & 0 & 0 & 0.0694 & 0 & 0 & 0 & 0 & 0 \\ 0 & 0 & 0 & 0 & 0 & 0.0786 & 0 & 0 & 0 & 0 \\ 0 & 0 & 0 & 0 & 0 & 0 & 0.1571 & 0 & 0 & 0 \\ 0 & 0 & 0 & 0 & 0 & 0 & 0 & 0.3367 & 0 & 0 \\ 0 & 0 & 0 & 0 & 0 & 0 & 0 & 0 & 0.3491 & 0 \\ 0 & 0 & 0 & 0 & 0 & 0 & 0 & 0 & 0 & 0.0786 \end{bmatrix}$$

$$\psi = \begin{bmatrix} 0.0598 & 0 & 0 & 0 & 0 & 0 & 0 & 0 & 0 & 0 \\ 0 & 0.0956 & 0 & 0 & 0 & 0 & 0 & 0 & 0 & 0 \\ 0 & 0 & 0.4781 & 0 & 0 & 0 & 0 & 0 & 0 & 0 \\ 0 & 0 & 0 & 0.3187 & 0 & 0 & 0 & 0 & 0 & 0 \\ 0 & 0 & 0 & 0 & 0.0478 & 0 & 0 & 0 & 0 & 0 \\ 0 & 0 & 0 & 0 & 0 & 0.0614 & 0 & 0 & 0 & 0 \\ 0 & 0 & 0 & 0 & 0 & 0 & 0.307 & 0 & 0 & 0 \\ 0 & 0 & 0 & 0 & 0 & 0 & 0 & 0.2632 & 0 & 0 \\ 0 & 0 & 0 & 0 & 0 & 0 & 0 & 0 & 0.307 & 0 \\ 0 & 0 & 0 & 0 & 0 & 0 & 0 & 0 & 0 & 0.0614 \end{bmatrix}$$

then we have $\lambda_{\max}(H_1^T H_1) = 0.0012$ and $\lambda_{\min}(\Xi H_1 + H_1^T \Xi) = 1.6455 \times 10^{-4}$ with system matrix $H_1 = \text{diag}\left\{2L_1^v + B_1^v, 2L_1^\omega + B_1^\omega\right\}$.

Similarly, based on $L_1 = \text{diag}\left\{L_1^P, L_1^Q\right\}$, we calculate that

$$\lambda_3(\psi L_1 + L_1^T \psi) = 0.0044 \tag{4.52}$$

$$\lambda_{\max}(\psi L_1 + L_1^T \psi) = 0.0022 \tag{4.53}$$

$$\lambda_{\max}(L_1^T L_1) = 0.0440 \tag{4.54}$$

In this way, by setting parameters $d = \hat{d} = 0.08$ and $d_1 = \hat{d}_1 = 0.1$ in Theorems 3.1 and 3.2, the delay-dependent conditions are verified as

$$d < \frac{(1 - d_1)\pi}{2\max_{i=2,\cdots,N}\left|\lambda_i\left(L_1^v\right)\right|} = 172.5882 \tag{4.55}$$

$$\tau_k(t) < 0.0688 < d < \frac{(1 - d_1)\lambda_{\min}\left(\Xi H_1 + H_1^T \Xi\right)}{\lambda_{\max}\left(H_1^T H_1\right) + (1 - d_1)\lambda_{\max}\left(\Xi H_1 H_1^T \Xi\right)} = 0.1271 \tag{4.56}$$

$$T_k(t) < 0.075 < \hat{d} < \frac{\left(1 - \hat{d}_1\right)\lambda_3\left(\psi L_1 + L_1^T \psi\right)}{\lambda_{\max}\left(L_1^T L_1\right) + \left(1 - \hat{d}_1\right)\lambda_{\max}\left(\psi L_1 L_1^T \psi\right)} = 0.0857 \tag{4.57}$$

The proof is similar for the other switching directed graphs in Fig. 4.3.

4.2.1 *Performance Evaluation*

Fixed Network Topology

As shown in Figs. 4.4 and 4.5, the evolution curves of voltage/frequency, reactive/active power and the output ratios is illustrated under the condition of no delay.

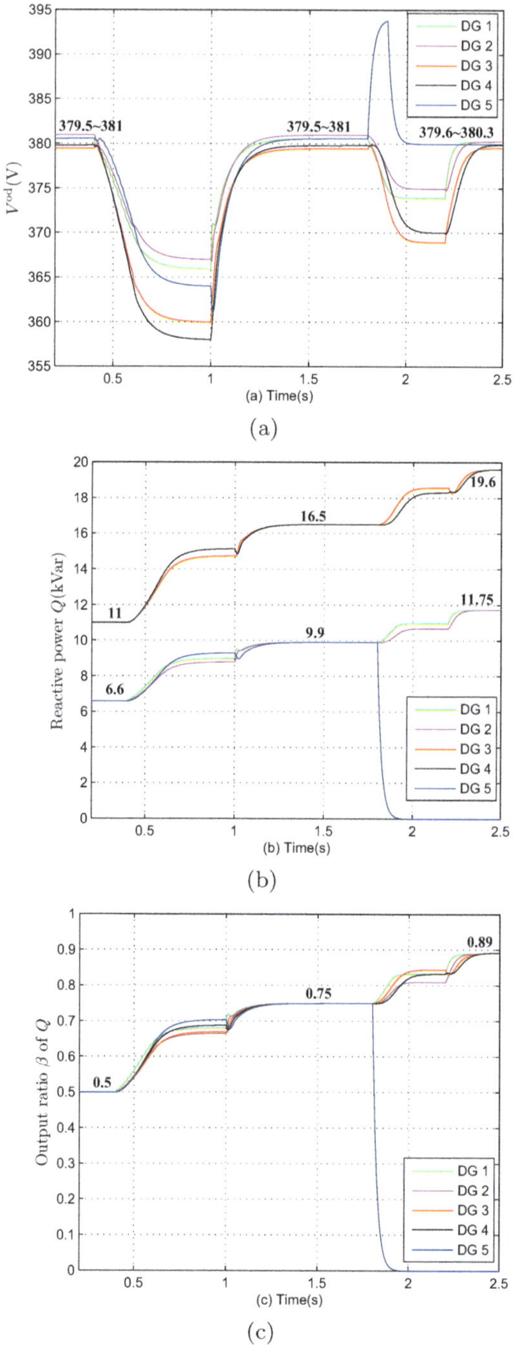

Fig. 4.4 The state evolution processes of voltage, reactive power and its output ratio without delays under $\hat{G}_1$ (©2024 IEEE. Reprinted, with permission, from IEEE Transactions on Smart Grid [189])

Fig. 4.5 The state evolution processes of frequency, active power and its output ratio without delays under $\hat{G}_1$ (©2024 IEEE. Reprinted, with permission, from IEEE Transactions on Smart Grid [189])

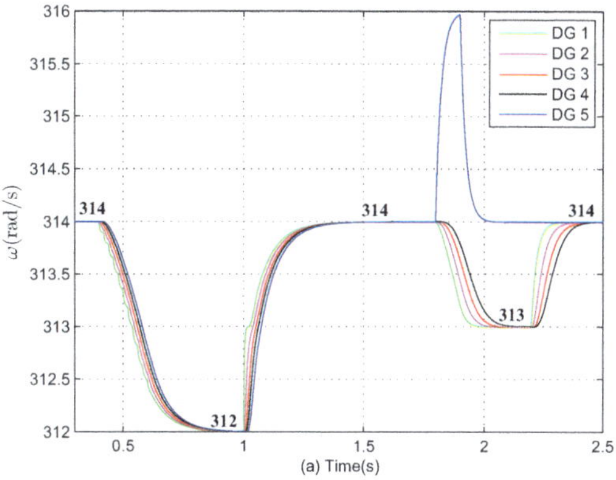

(a)

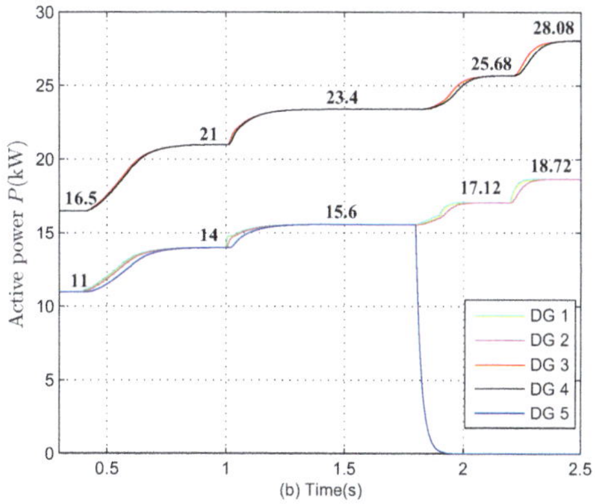

(b)

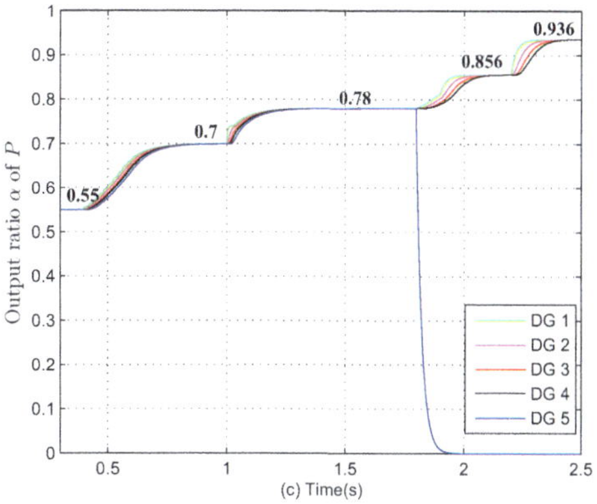

(c)

At $t = 0.4$ s, the microgrid breaks away from the main grid and enters the island operation mode.

As shown in Figs. 4.4a and 4.5a, the direct term of the output voltage v^{od} and frequency ω of each DER agent in the microgrid all vary in the direction lower than the reference value $v^{\mathrm{DES}} = 380$ V and $\omega^{\mathrm{DES}} = 314$ rad/s after entering the island operation. After the droop control technique is adopted, as for DER agents' output voltage DER_i agent,the direct terms converge to the range of 358 V to 367 V, while quadrature terms converge to 0. At the same time, the output frequency ω_i of DER agent converges to 312 rad/s because of the fact that frequency is a global variable and voltage is a local variable.

After adding the distributed intelligent cooperative protocol in the secondary control level at $t = 1$ s, the output voltage and output frequency gradually converge to the reference value at $t = 0$ s. Due to the existence of precise reactive power sharing, the output voltage of each DER agent fails to precisely converge to the previous reference value. On the basis of this fact, both the output voltage and the output frequency are slowly eliminating the steady-state error caused by the droop control process in Fig. 4.4a. Instead, we find a set of weights $\mu = (0.0694, 0.1387, 0.2601, 0.4624, 0.0694)^T$ that guarantees that the weighted average value of all DER agents converges to a preset reference value 380 V.

Finally, the process of voltage and frequency restoration after single DER agent disconnect is shown. At time $t = 1.8$ s, remove DER_5 agent from the communication topology and disconnect links 4-5 and 5-1 between it and its neighbors DER_1 and DER_4. When droop control is introduced at $t = 2.4$ s, the new digraph still has strong connectivity despite the disconnection of DER_5 agent, so the previous theory is still valid. Therefore, the voltage and frequency of the remaining DER agents after DER_5 agent disconnection can still be restored to the reference value. At the same time, the voltage and frequency of DER_5 agent are also quickly restored to the reference value under the action of local droop control and trigger protection strategy.

The state evolution process of reactive/active power and their ratio is shown in Figs. 4.4b, c and 4.5b, c. When the microgrid enters the island mode at $t = 0.4$ s and DER_5 agent is disconnected at $t = 1.8$ s, the active/reactive power and their ratio converge to a new value. Comparing Figs. 4.4 and 4.5, droop control causes a deviation in the reactive power of each DER agent while the active power can be precisely shared. This result is because the frequency of DER agents is a global variable but the voltage is not, the same as in the previous analysis. After the distributed intelligent cooperative protocol in the secondary control level is adopted at $t = 1$ s and $t = 2.4$ s, the reactive/active power of all DER agents can be precisely shared, which verifies the effectiveness of this strategy. Finally,the above results in terms of the output ratio of reactive/active power are also verified in Figs. 4.4c and 4.5c.

As shown in Figs. 4.6 and 4.7, the state evolution process of reactive/active power, voltage/frequency and their output ratio with communication delay is illustrated. By comparing Figs. 4.4, 4.5 and 4.6, 4.7, it is shown that delay will slow down the convergence rate of sag control and distributed intelligent cooperative

Fig. 4.6 The state evolution processes of voltage, reactive power and its output ratio with delays under $\hat{G}_1$ (©2024 IEEE. Reprinted, with permission, from IEEE Transactions on Smart Grid [189])

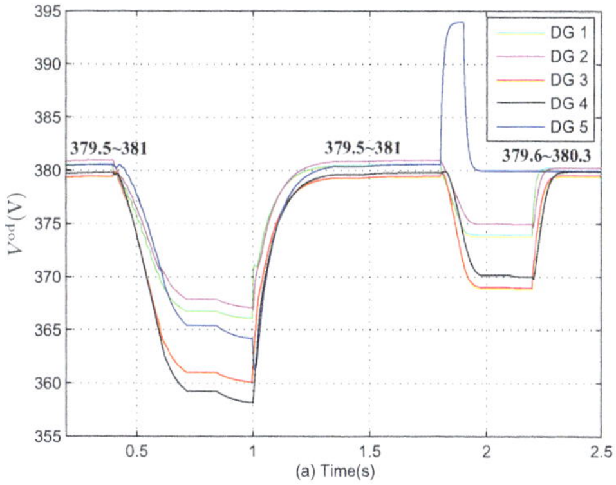

(a)

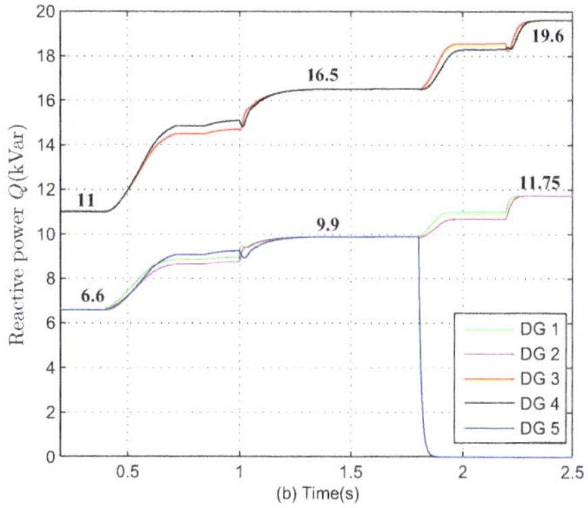

(b)

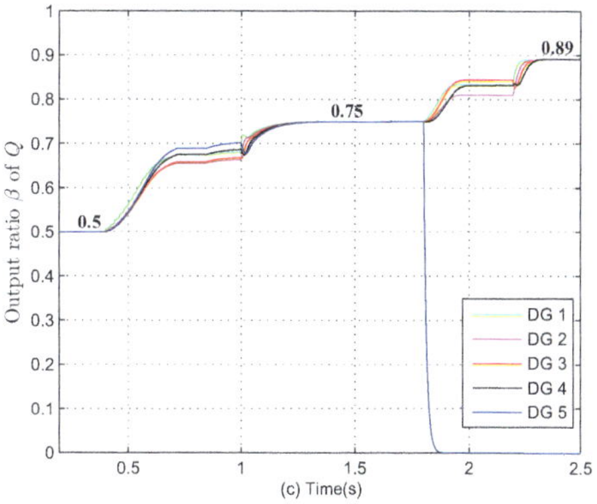

(c)

Fig. 4.7 The state evolution processes of frequency, active power and its output ratio with delays under $\hat{G}_1$ (©2024 IEEE. Reprinted, with permission, from IEEE Transactions on Smart Grid [189])

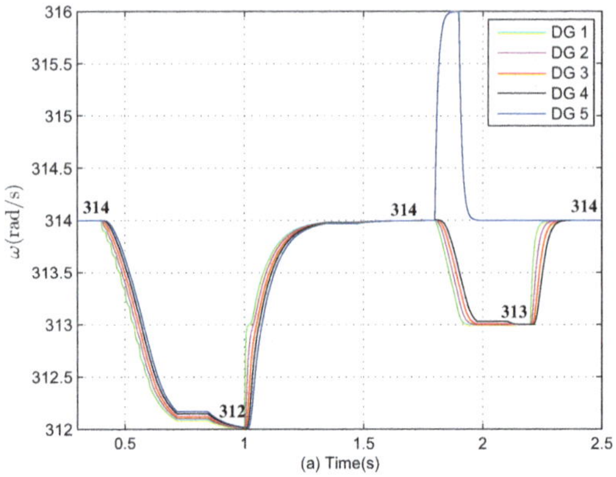

(a)

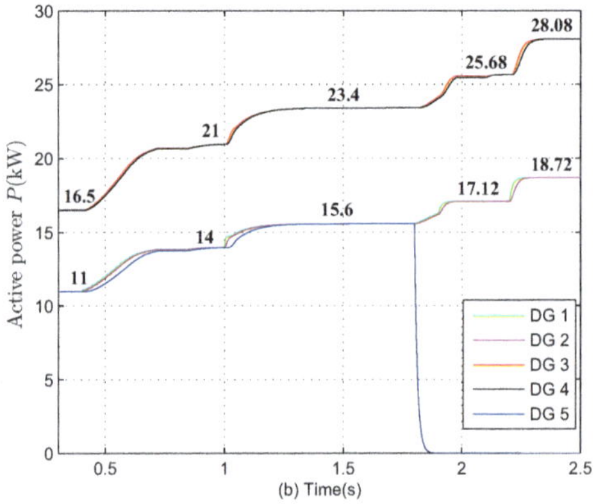

(b)

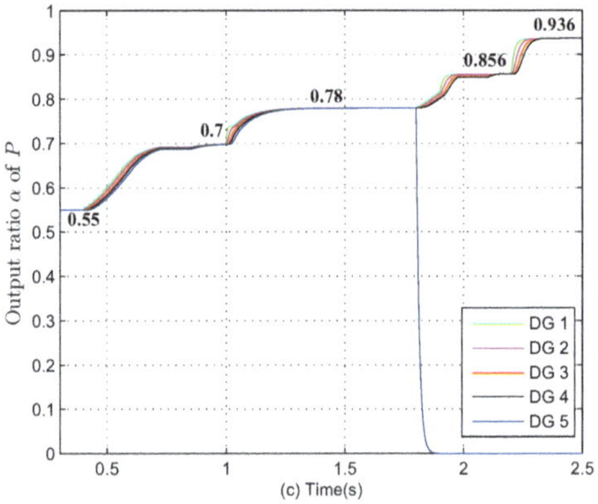

(c)

protocol. Specifically, the original settling time of the droop control is delayed from $t = 0.8$ s to $t = 1$ s, while the convergence time of the distributed intelligent cooperative protocol is delayed from $t = 1.4$ s to $t = 1.6$ s.

Dynamic Network Topology

Consider the case of dynamical network topology of a digraph, which is shown as $\hat{G}_1$-$\hat{G}_5$ in Fig. 4.3. The initial topology is $\hat{G}_1$, which is switched after $\Delta t = 0.001$ s according to the switching sequence $\{G_1, \cdots, G_5, G_1, \cdots, G_5\}$. All other parameters are the same as the fixed topology analyzed earlier.

Figures 4.8 and 4.9 illustrates the state evolution curves of voltage/frequency, reactive power/active power and their ratio with communication delay in dynamical network topology. Compared with Figs. 4.6 and 4.7, it is shown that there is oscillation in the process of convergence in Figs. 4.8 and 4.9, because the control system contains frequent switching signals. Next, we choose the delay functions $\tau_k(t) = \frac{1+sin(t+k)}{10}$ and $T_k(t) = \frac{1+sin(2t+2k)}{10}$ that do not satisfy the constraints Theorems 3.1–3.2, and prove the validity of proposed droop-based distributed intelligent cooperative protocol from the opposite side. As shown in Figs. 4.10 and 4.11, voltage/frequency, reactive power/active power, and their ratios cannot converge to reference values.

4.2.2 *Model Verification and Time Delay Analysis*

In the distributed droop-based control strategy shown in Fig. 4.1, ω^{DES} and v^{DES} are the inputs while ω, v^{od}, P and Q are the outputs. In order to further verify the validity of the proposed model, we compare the responses of the model through the simulation results in Fig. 4.12, which indicates all the step response curves fit well.

By analyzing the inequality conditions (4.16) and (4.31) obtained from Theorems 3.1–3.2, we summarize the main factors affecting the stability of the system. Besides the processing delay, it also includes the topology structure and the weight of the communication network link. Due to space limitation, we only focus on the influence of different delays on the system stability in fixed topology $\hat{G}_1$.

From the previous analysis, the upper bound and change rate of the delay $d(\hat{d})$ and $d_1(\hat{d}_1)$ will affect the control performance of the system. In order to study the sensitivity of these parameters in the system, we set a large parameter $d(\hat{d})$ while fixing parameter $d_1(\hat{d}_1)$ in the simulation, as shown in Fig. 4.13. In Fig. 4.14, while fixing parameter $d(\hat{d})$, a large parameter $d_1(\hat{d}_1)$ is set. Here the larger values of the parameters $d(\hat{d})$ and $d_1(\hat{d}_1)$ are taken close to the thresholds of the inequality (4.16) and (4.31). After comparison, the control effect in Fig. 4.13 is better than that in Fig. 4.14, but both results are worse than those shown in Fig. 4.12. This is because the upper bounds of the parameters in Fig. 4.12 are well below the thresholds.

Fig. 4.8 The state evolution processes of voltage, reactive power and its output ratio with delays under dynamical network digraph topology (©2024 IEEE. Reprinted, with permission, from IEEE Transactions on Smart Grid [189])

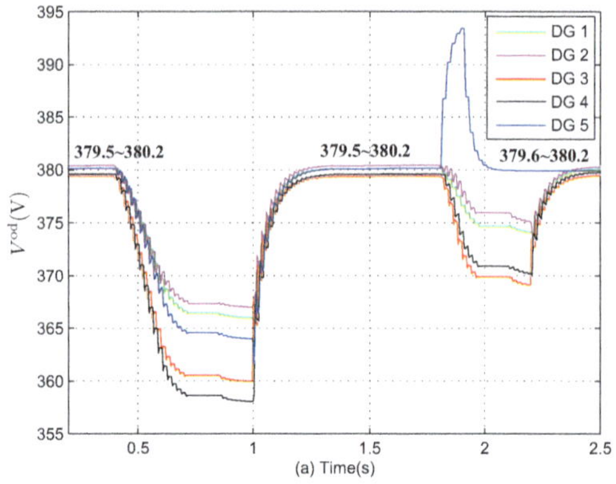

(a)

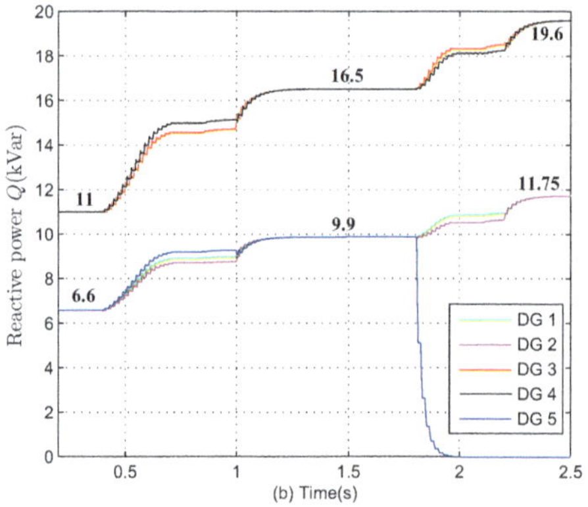

(b)

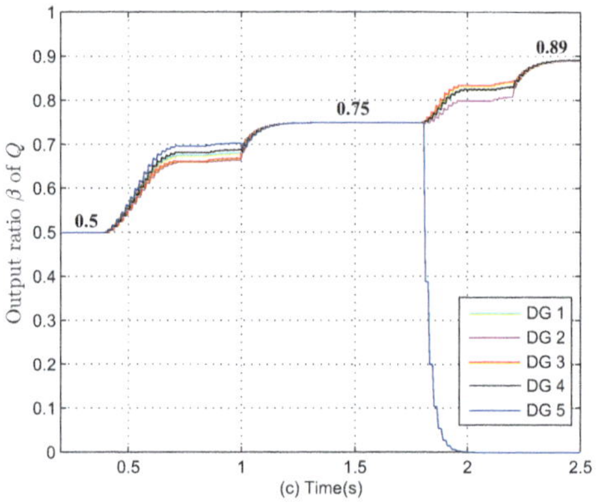

(c)

Fig. 4.9 The state evolution processes of frequency, active power and its output ratio with delays under dynamical network digraph topology (©2024 IEEE. Reprinted, with permission, from IEEE Transactions on Smart Grid [189])

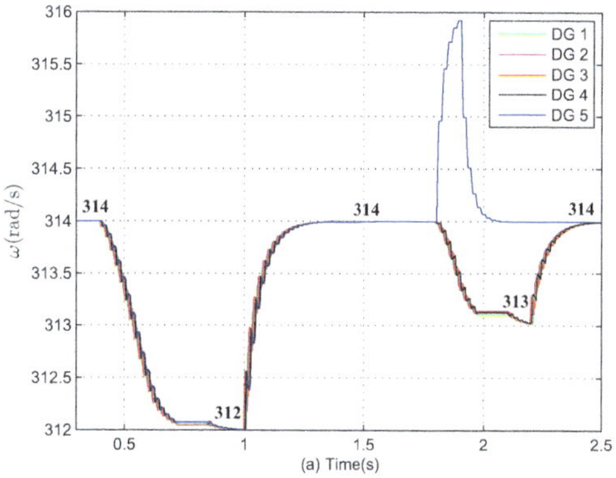

(a)

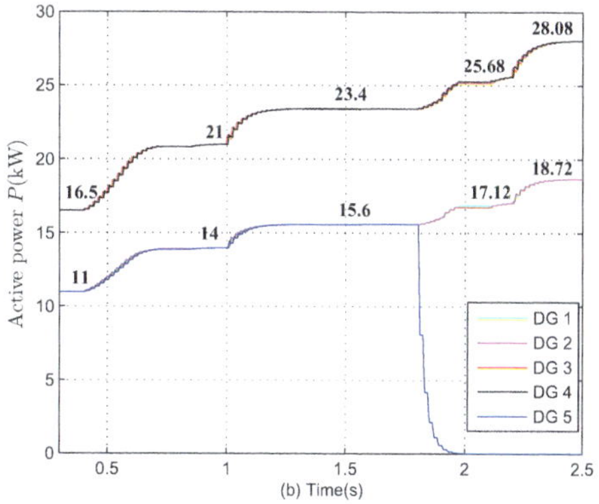

(b)

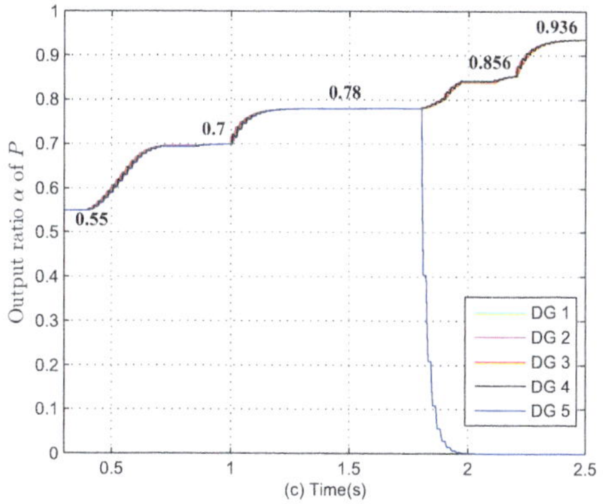

(c)

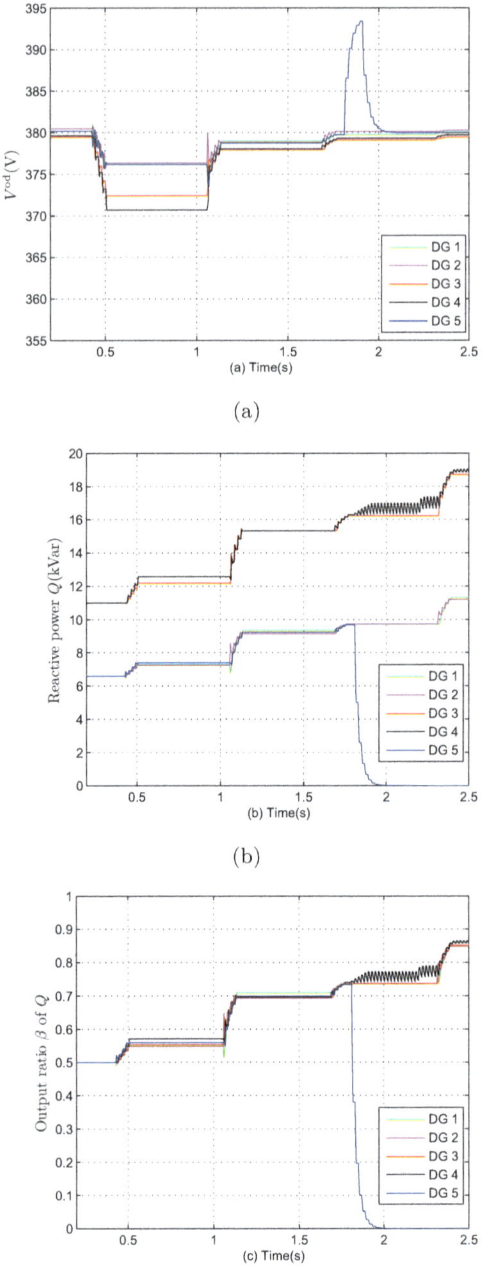

(a)

(b)

(c)

Fig. 4.10 The state evolution processes of voltage, reactive power and its output ratio with large delays under dynamical network digraph topology (©2024 IEEE. Reprinted, with permission, from IEEE Transactions on Smart Grid [189])

Fig. 4.11 The state evolution processes of frequency, active power and its output ratio with large delays under dynamical network digraph topology (©2024 IEEE. Reprinted, with permission, from IEEE Transactions on Smart Grid [189])

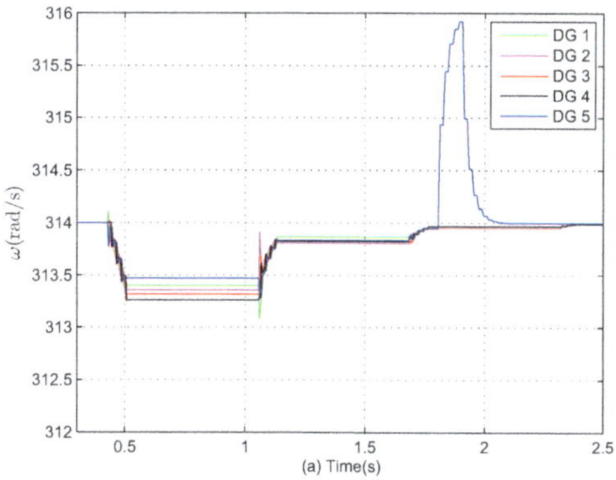

(a)

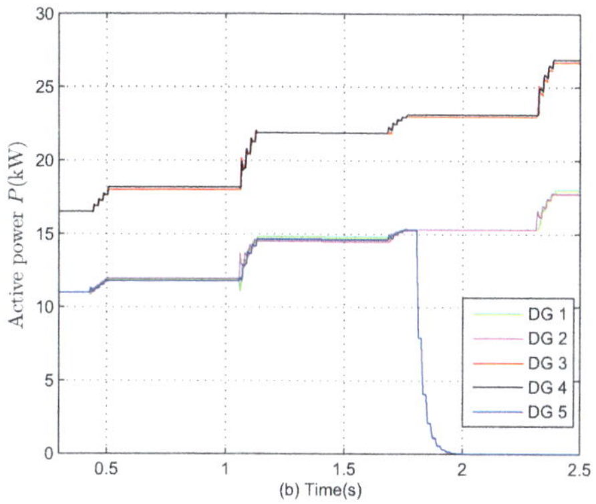

(b)

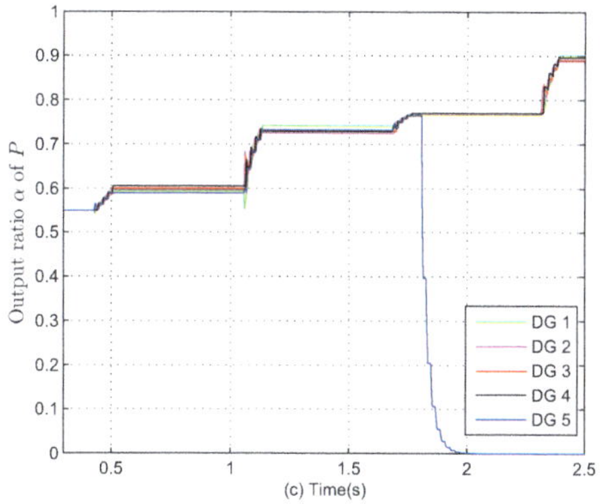

(c)

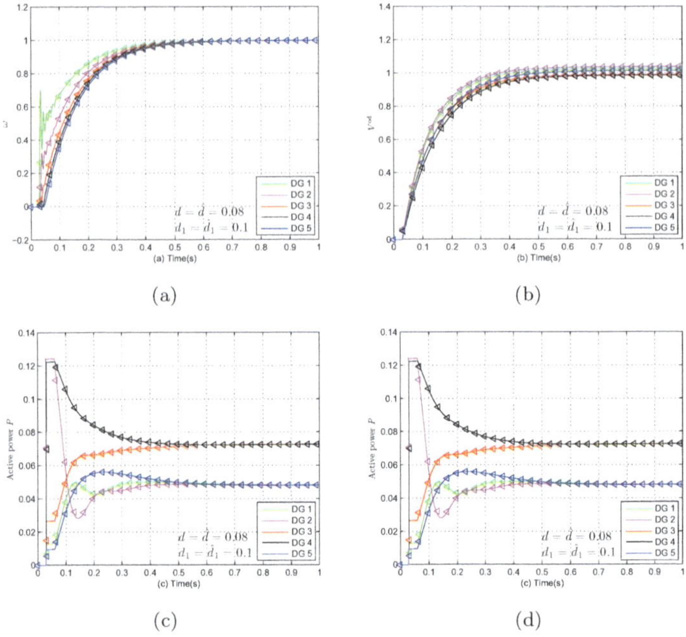

Fig. 4.12 The step response of controlled system with $\tau_k(t) = \frac{1+0.1sin(t+k)}{16}$ and $T_k(t) = \frac{1+0.2sin(2t+2k)}{16}$ ($\triangleleft$ represents the response of the state space equations) (©2024 IEEE. Reprinted, with permission, from IEEE Transactions on Smart Grid [189])

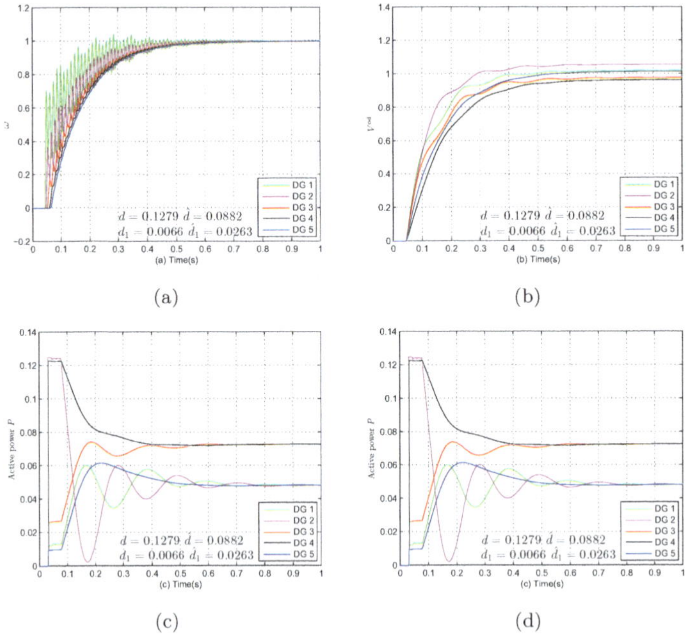

Fig. 4.13 The step response of controlled system with $\tau_k(t) = \frac{1.95+0.1sin(t+k)}{16}$ and $T_k(t) = \frac{1.2+0.2sin(2t+2k)}{16}$ (©2024 IEEE. Reprinted, with permission, from IEEE Transactions on Smart Grid [189])

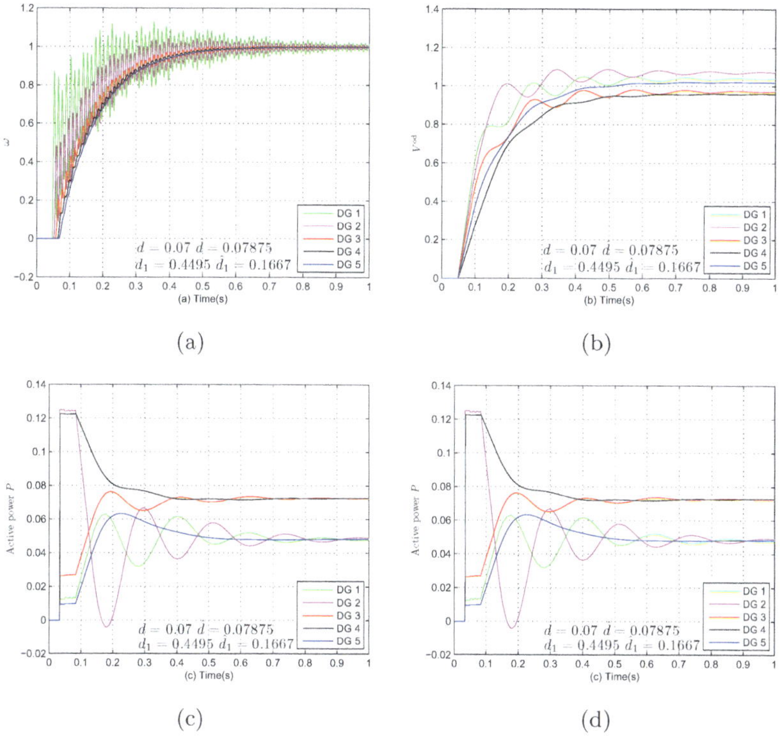

Fig. 4.14 The step response of controlled system with $\tau_k(t) = \frac{1.1sin(6.5t+k)}{16}$ and $T_k(t) = \frac{1.2sin(2t+2k)}{16}$ (©2024 IEEE. Reprinted, with permission, from IEEE Transactions on Smart Grid [189])

As shown in Figs. 4.15 and 4.16, continue increasing the values of $d(\hat{d})$ and $d_1(\hat{d}_1)$ so that they exceed the upper bound. As a result, this causes the system dynamics to become unstable, illustrating the validity of the previously calculated threshold from the opposite side. Compared with Figs. 4.15 and 4.16, under the condition of fixed topology and link weight, the delay change rate has a greater impact on system stability. In summary, the system dynamics have strong robustness to large constant delay functions, but it is difficult to cope with small delay functions with high change rates.

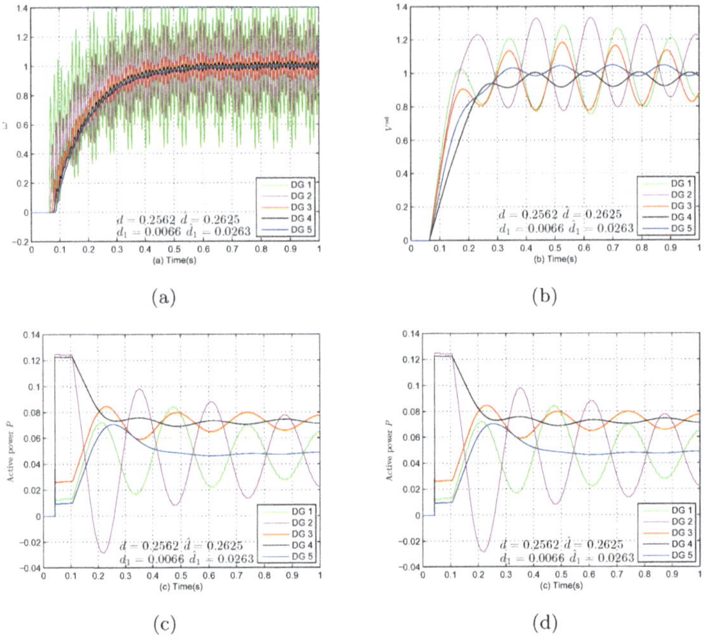

Fig. 4.15 The step response of controlled system with $\tau_k(t) = \frac{4+0.1sin(t+k)}{16}$ and $T_k(t) = \frac{4+0.2sin(2t+2k)}{16}$ (©2024 IEEE. Reprinted, with permission, from IEEE Transactions on Smart Grid [189])

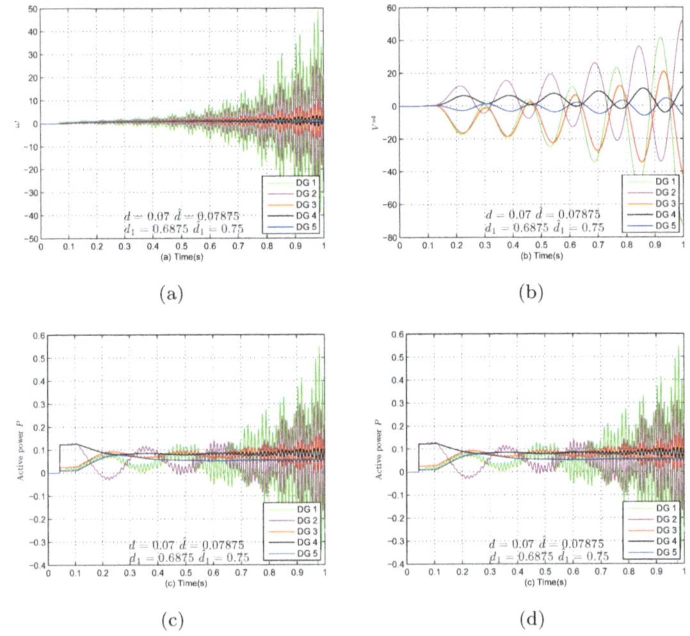

Fig. 4.16 The step response of controlled system with $\tau_k(t) = \frac{1.1sin(10t+k)}{16}$ and $T_k(t) = \frac{1.2sin(10t+2k)}{16}$ (©2024 IEEE. Reprinted, with permission, from IEEE Transactions on Smart Grid [189])

4.3 Summary

This chapter proposes a droop-based distributed cooperative control under the condition of non-uniform delay. By adopting this control strategy, the voltage/frequency and reactive/active power can be convergent to the reference value. This method is suitable for fixed and switching topologies, as long as the delay upper bound does not exceed the threshold and the topology has strong connectivity. Simulation results show that the method is effective.

Chapter 5
Discrete-Time Distributed Cooperative Intelligence for AC Microgrids with Uncertain Communication Links

This chapter discusses the robust discrete-time distributed intelligent cooperative protocol for AC microgrids in a distribution sparse network with uncertain communication links. To achieve the voltage regulation, frequency restoration as well as accurate power sharing for all the distributed energy resources, the discrete-time distributed intelligent cooperative protocol is designed for AC microgrids based on the iterative learning mechanics. The discussed controller appropriately responds to the uncertain communication links through an iterative operation manner, with only the requirement of accessing limited information of the system parameters. And the input of the proposed distributed intelligent cooperative protocol are simply updated at the end of each iteration operation, such that one DER agent only needs to communicate with their neighbors intermittently. Therefore, the communication cost can be greatly reduced. And some sufficient conditions on the system stability and robustness to the uncertainties are also derived by using the tools of direct Lyapunov method, algebraic graph theory, and linear matrix inequality theory. Moreover, the desired control objective can also be guaranteed even if all DER agents are subject to internal uncertainties and external noises including initial voltage and/or frequency resetting errors and measurement disturbances, which then improves the system reliability and robustness.

Different from traditional centralized control for microgrids with many deficiencies, which include single point of failure, low scalability and flexibility, high dependency on communication infrastructure and poor resilience. The distributed structure has the low costs generated by the sparse communication network, simplicity for the future extension of microgrids [191, 219, 220], and provides a practically feasible and highly computational efficiency solution for distributed intelligent cooperative protocol design. Based on the structure, some distributed intelligent cooperative protocol in the secondary control level methods have been proposed to achieve voltage regulation [191, 221], frequency restoration [40, 222], reactive sharing [136, 223], and imbalance voltage compensation [77], etc. Specially, a distributed secondary control approach for voltage regulation was investigated in

J. Lai, X. Lu, *Distributed Cooperative Intelligence in Microgrid Control*,
https://doi.org/10.1007/978-981-95-6884-0_5

[191] based on the distributed cooperative control of multi-agent systems [224, 225] and the feedback linearization technique with high gains. Unfortunately, the high gain design leads to the performance degradation of the microgrid control system, and even has a negative affect on the overall stability of microgrids. Then, a distributed secondary control method was proposed in [77] to guarantees the voltage regulation in a finite time. In order to ensure the achievement of the power output regulation of an AC microgrid with uncertain communication topologies, a pinning-based distributed secondary control method scheme was proposed in [226], which is of benefit to meet the common requirements of line switches and plug-and-play operation in microgrids. Besides, a distributed secondary control approach was proposed in [40] to regulate the voltage, frequency, and reactive power for the AC microgrid. However, each local DER agent under this control method is required to exchange information with all the other DER agents in the microgrid, which results almost the same communication costs as the centralized control approach. Moreover, in order to obtain the tradeoff between voltage regulation and accurate reactive power sharing, the information exchange among neighboring DER agents must be stable guarantee [193]. However, in the aforementioned literature, it is generally assumed that all DER agents can access the system parameters and neighboring DER agents' information accurately, and the unavoidable communication disturbances and noises are ignored.

It is difficult for each DER agent to get the accurate information from its neighbors due to the impact of numerical errors [135], component mismatches, transient power oscillation, high frequency disturbances and noises [136, 227], which impacts on the stability of low-voltage microgrids are more severe than that of high-voltage ones [136, 137]. Moreover, DER agents inevitably be affected by the perturbations in the process of information exchange [77, 190, 216]. Summarized the above, the impacts of communication disturbances and noises on the microgrid system must be taken into account when we design distributed secondary control schemes for microgrids.

This chapter aims to propose a novel robust distributed secondary control scheme for islanded microgrids with uncertain communication links. In detail, two distributed secondary controllers with discrete-time control inputs are designed on local DER agents for the final implementation of voltage regulation and frequency restoration and as well as the accurate active power sharing. Since the designed control inputs are updated at the end of each iteration interval, the DER agent under the proposed distributed secondary control scheme only needs to perform merely local measurements and communicate with its neighbors intermittently. Stability analysis of the islanded microgrid system under the proposed distributed secondary control scheme is given, and the corresponding sufficient conditions are also derived. Moreover, the information exchanges among DER agents under the proposed distributed secondary control scheme are in the discrete form with longer intervals, thus, the communication costs can be greatly reduced, which makes our results essentially different from the existing methods that with the continuous time communications [191, 226].

5.1 Problem Statement

5.1.1 Problem Formulation

In order to maintain the stability of the voltage and frequency in microgrid, the primary control is utilized as the power control loop for the ith DER agent [197]:

$$\begin{cases} v_i = v_i^{\text{nom}} - n_i^Q Q_i \\ \omega_i = \omega_i^{\text{nom}} - m_i^P P_i \end{cases} \tag{5.1}$$

where v_i and ω_i are the actual voltage magnitude and frequency, respectively; v_i^{nom} and ω_i^{nom} are the nominal set points of voltage and frequency, which are determined at the secondary control layer; m_i^P and n_i^Q are the associated droop coefficients, respectively; P_i and Q_i are the measured output active and reactive power after the low-pass filters, which can be calculated by

$$\begin{cases} P_i = \dfrac{\omega_i^{\text{c}}}{\omega_i^{\text{c}} + s} P_i^{\text{m}} = \dfrac{\omega_i^{\text{c}}}{\omega_i^{\text{c}} + s} \left(v_i^{\text{od}} i_i^{\text{od}} + v_i^{\text{oq}} i_i^{\text{oq}} \right) \\ Q_i = \dfrac{\omega_i^{\text{c}}}{\omega_i^{\text{c}} + s} Q_i^{\text{m}} = \dfrac{\omega_i^{\text{c}}}{\omega_i^{\text{c}} + s} \left(v_i^{\text{od}} i_i^{\text{oq}} - v_i^{\text{oq}} i_i^{\text{od}} \right) \end{cases} \tag{5.2}$$

where ω_i^{c} is the cutoff frequency of the first-order low-pass filter installed in the ith DER agent's power control loop; v_i^{od} and v_i^{oq} are the direct and quadrature components of output voltage, respectively; i_i^{od} and i_i^{oq} are the direct and quadrature components of operating frequency, respectively; P_i^{m} and Q_i^{m} are the instantaneous active and reactive power components calculated from the measured output voltage and output current [190].

In the primary voltage control strategy, we choose the output voltage reference is projected on the d-axis of reference frame of the ith DER agent, the droop control principle (5.1) for the ith DER agent can be written as

$$\begin{cases} v_i^{\text{od}} = v_i^{\text{nom}} - n_i^Q Q_i \\ v_i^{\text{oq}} = 0 \\ \omega_i = \omega_i^{\text{nom}} - m_i^P P_i \end{cases} \tag{5.3}$$

Since the output voltage amplitude of DER$_i$ agent is $v_i = \sqrt{(v_i^{\text{od}})^2 + (v_i^{\text{oq}})^2}$, the voltage regulation objective $v_i \to v^{\text{DES}}$ can be achieved by choosing a reasonable voltage nominal set point v_i^{nom} such that v_i^{od} can be restored to $v_i \to v^{\text{DES}}$.

5.1.2 *Uncertain Communication Link*

For the uncertain communication link, it can be assumed that each nonzero elements of the weighted adjacency matrix a_{ij} satisfies $0 < \underline{a}_{ij} \leq a_{ij} \leq \overline{a}_{ij}$, where $\underline{a}_{ij}$ and $\overline{a}_{ij}$ are the nonnative bounds. Denote $\underline{A}$ and $\overline{A}$ are the corresponding lower bound matrix and upper bound matrix, which respectively correspond to the digraph $G(\underline{A})$ and $G(\overline{A})$. It is obvious that $G(\underline{A})$ is a spanning sub-graph of $G(\overline{A})$, and the associated neighbor sets of the ith DER agent satisfies $\underline{\mathcal{N}_i} \subseteq \mathcal{N}_i \subseteq \overline{\mathcal{N}_i}$.

The voltage and frequency as well as active power may possess heterogeneous network structures and weights, the corresponding digraphs can be represented by $\hat{G}(A^v)$, $\hat{G}(A^\omega)$ and $\hat{G}(A^P)$.

5.1.3 *Useful Lemma*

Lemma 5.6 ([228]) *Let λ be an eigenvalue of matrix A, and α and β be two vectors such that:*

(1) $A\alpha = \lambda\alpha$;
(2) $A^T\beta = \lambda\beta$;
(3) $\alpha^T\beta = 1$;
(4) $|\lambda| = \rho(\lambda) > 0$;
(5) $|\lambda| = \rho(\lambda) > 0$.

Define $\hbar = \alpha\beta^T$, and the $(\lambda^{-1}A)^\zeta = \hbar + (\lambda^{-1}A - \hbar)^\zeta \to \hbar$ as $\zeta \to \infty$. Then, some necessary notifications are given below. For any vectors $x, y \in R^n$, we denote

$$x \odot y = [x_1y_1, \cdots, x_ny_n]^T \in R^n \tag{5.4}$$

where $x = [x_1, \cdots, x_n]^T$ and $y = [y_1, \cdots, y_n]^T$.

5.2 Distributed Cooperative Intelligence with Discrete Control Inputs

In practice, the continuous variables v_i^{od}, ω_i, P_i and Q_i must be discredited to prompt the realization of the distributed intelligent cooperative protocol can be implemented in the actual microgrid. Different from the conventional distributed intelligent cooperative protocol in the secondary control level with the continuous-time control dynamics [191, 226], this section will proposed a new distributed intelligent cooperative protocol to achieve the voltage regulation and frequency

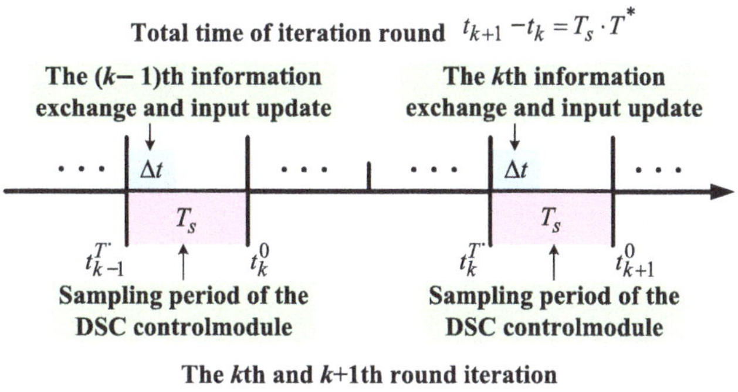

Fig. 5.1 Time diagram of the proposed discrete-time distributed intelligent cooperative protocol

restoration as well as active power sharing for AC microgrid through the discrete-time control inputs.

Firstly, the system (5.3) is transformed into a discrete-time system with a sampling period T_s. Then the continuous time period can be discretized into a finite-time sequence of nonempty and bounded intervals, $[t_k, t_{k+1})$, where $t_0 = 0$ and $k = 0, 1, 2, \cdots$ representing the kth round iteration index, as shown in Fig. 5.1. It should be pointed out that there is a sequence of non-overlapping sub-intervals $[t_k^0, t_k^1), [t_k^1, t_k^2), \cdots, [t_k^{T^*-1}, t_k^{T^*})$with $t_k^0 = t_k$, $t_k^{T^*} = t_{k+1}$, which satisfy $t_{k+1} - t_k = T^*$ and $t_k^{\ell+1} - t_k^{\ell} = T_s$ during each interval $[t_k, t_{k+1})$. Then, the control input will be updated at the end of each iterative process, i.e. $[t_k^{T^*}, t_k^{T^*} + \Delta t$, where $\Delta t \ll T_s$. For simplicity, the time period T^* is regarded as the step length of the control input update, and $t_k^{T^*} = k \cdot T^* \cdot T_s$ is the terminal time of the system state output.

Then, the discrete-time control system in the secondary control layer can be expressed as

$$v_i^{\text{od}}(t_k^{\ell+1}) = v_i^{\text{od}}(t_k^{\ell}) + u_{i,k}^{v}(t_k^{\ell}) \tag{5.5}$$

$$\omega_i(t_k^{\ell+1}) = \omega_i(t_k^{\ell}) + u_{i,k}^{\omega}(t_k^{\ell}) \tag{5.6}$$

$$P_i(t_k^{\ell+1}) = P_i(t_k^{\ell}) + u_{i,k}^{P}(t_k^{\ell}) \tag{5.7}$$

$$Q_i(t_k^{\ell+1}) = Q_i(t_k^{\ell}) + u_{i,k}^{Q}(t_k^{\ell}) \tag{5.8}$$

where $u_{i,k}^{v}(t_k^{\ell})$, $u_{i,k}^{\omega}(t_k^{\ell})$, $u_{i,k}^{P}(t_k^{\ell})$ and $u_{i,k}^{Q}(t_k^{\ell})$ are the distributed auxiliary controllers for voltage regulation, frequency restoration, active and reactive power sharing;

According to the droop control algorithm in (5.1), the nominal set point for the ith DER agent can be calculated by

$$\begin{cases} v_i^{\text{nom}}(t_k^{\ell+1}) = v_i^{\text{od}}(t_k^{\ell+1}) + n_i^Q Q_i(t_k^{\ell+1}) \\ \qquad = v_i^{\text{od}}(t_k^{\ell}) + n_i^Q Q_i(t_k^{\ell}) + u_{i,k}^v + n_i^Q u_{i,k}^Q \\ \qquad = v_i^{\text{nom}}(t_k^{\ell}) + u_{i,k}^{v,Q} \\ \omega_i^{\text{nom}}(t_k^{\ell+1}) = \omega_i(t_k^{\ell+1}) + m_i^P P_i(t_k^{\ell+1}) \\ \qquad = \omega_i(t_k^{\ell}) + P_i^Q P_i(t_k^{\ell}) + u_{i,k}^{\omega} + m_i^P u_{i,k}^P \\ \qquad = \omega_i^{\text{nom}}(t_k^{\ell}) + u_{i,k}^{\omega,P} \end{cases} \tag{5.9}$$

where distributed auxiliary controllers of the nominal set point are respectively defined as follows

$$u_{i,k}^{v,Q} = u_{i,k}^v + n_i^Q u_{i,k}^Q \tag{5.10}$$

$$u_{i,k}^{\omega,P} = u_{i,k}^{\omega} + m_i^P u_{i,k}^P \tag{5.11}$$

Due to the inherent contradiction between the voltage regulation and reactive power sharing in AC microgrids, we mainly focus on the voltage regulation. Thus, the reactive power control inputs are updated from [190, 191] as

$$u_{i,k}^Q = -\omega_i^c Q_i(t_k^{\ell}) + \omega_c Q_i^{\text{m}}(t_k^{\ell}) \tag{5.12}$$

The control objectives of the secondary control layer are design the discrete-time controllers $u_{i,k}^v$, $u_{i,k}^{\omega}$to regulate the voltage and frequency of all DER agents can separately synchronize to their desired value and design the discrete-time controllers $u_{i,k}^P$ to achieve precise power sharing exactly at the terminal time $t_k^{T^*}$. The control objectives can be summarized as following:

$$\lim_{k\to\infty} v_i^{\text{od}}(t_k^{T^*}) = v^{\text{DES}} \tag{5.13}$$

$$\lim_{k\to\infty} \omega_i(t_k^{T^*}) = \omega^{\text{DES}} \tag{5.14}$$

$$\lim_{k\to\infty} \left(\frac{P_i(t_k^{T^*})}{P_{i,\max}} - \frac{P_j(t_k^{T^*})}{P_{j,\max}} \right) \tag{5.15}$$

for all $i \neq j = 1, 2, \cdots, N$, where $P_{i,\max}$ is the active power rating of the ith DER agent, v^{DES} and ω^{DES} are the constant desired values of voltage and frequency, respectively.

Summarize the above, the aforementioned control problems of voltage/frequency restoration and active power sharing can be transformed to a class of desired finite-time synchronization problem [228]. However, the current control protocols, stability analysis, and the control objective are completely different. From this perspective, the proposed discrete time distributed intelligent cooperative protocol is essentially different from the voltage/frequency restoration and active power sharing considered in [191, 226], and [40].

Recall Fig. 5.1, the during the kth iteration round, the information exchange among DER agents begins at $t = t_k^0$ and the control inputs $u_{i,k}^v$, $u_{i,k}^\omega$, and $u_{i,k}^P$ needs to be updated at $t = t_k^0 + \Delta t$ only once. Then, the microgrid system variables v_i^{od}, ω_i, and P_i will be updated at each discrete time $t = t_k^1, \cdots, t_{k\ell}, \cdots, t_k^{T*}$ with the invariant control inputs $u_{i,k}^v$, $u_{i,k}^\omega$, and $u_{i,k}^P$, respectively.

5.2.1 Voltage and Frequency Restoration

In this section, we will design the following discrete-time controllers $u_{i,k}^v$ and $u_{i,k}^\omega$ by incorporating the relative errors between the neighboring DER agents at terminal time $t_k^{T^*}$ to synchronize the voltage and frequency to their desired value.

$$
\begin{aligned}
u_{i,k}^v = & \sum_{j\in\mathcal{N}_i^v} c\gamma_{ij}^v a_{ij}^v \left[v_j^{\text{od}}(t_k^{T^*}) - v_i^{\text{od}}(t_k^{T^*}) \right] \\
& + c\gamma_{i0}^v a_{i0}^v \left[v^{\text{DES}} - v_i^{\text{od}}(t_k^{T^*}) \right]
\end{aligned} \tag{5.16}
$$

$$
\begin{aligned}
u_{i,k}^v = & \sum_{j\in\mathcal{N}_i^v} c\gamma_{ij}^v a_{ij}^v \left[v_j^{\text{od}}(t_k^{T^*}) - v_i^{\text{od}}(t_k^{T^*}) \right] \\
& + c\gamma_{i0}^v a_{i0}^v \left[v^{\text{DES}} - v_i^{\text{od}}(t_k^{T^*}) \right]
\end{aligned} \tag{5.17}
$$

for $i = 1, 2, \cdots, N$, where c is a positive constant representing the coupling strength of the cyber network, γ_{ij}^v and γ_{ij}^ω ($i = 1, 2, \cdots, N$ and $j = 0, 1, \cdots, N$) are the control gain determining by iteration.

The control objective (5.13) can be achieved asymptotically under the distributed voltage and frequency controllers (5.16) if their certain conditions are satisfied. Then we will derive the conditions.

We first denote the weighted adjacency matrices of voltage and frequency are

$$
A^v = (a_{ij}^v)_{N\times N} \in [\underline{A}^v, \overline{A}^v] \tag{5.18}
$$

$$
A^\omega = (a_{ij}^\omega)_{N\times N} \in [\underline{A}^\omega, \overline{A}^\omega] \tag{5.19}
$$

And the corresponding leader adjacency matrices are defined as follows

$$B^{v}=(a_{i0}^{v})_{N\times N}\in[\underline{B}^{v},\overline{B}^{v}] \tag{5.20}$$

$$B^{\omega}=(a_{i0}^{\omega})_{N\times N}\in[\underline{B}^{\omega},\overline{B}^{\omega}] \tag{5.21}$$

Furthermore, the associated control gain matrices are

$$\Gamma^{v}=(\gamma_{ij}^{v})_{N\times N} \tag{5.22}$$

$$\Gamma^{\omega}=(\gamma_{ij}^{\omega})_{N\times N} \tag{5.23}$$

$$\Xi^{v}=\mathrm{diag}\{\gamma_{10}^{v},\cdots,\gamma_{N0}^{v}\} \tag{5.24}$$

$$\Xi^{\omega}=\mathrm{diag}\{\gamma_{10}^{\omega},\cdots,\gamma_{N0}^{\omega}\} \tag{5.25}$$

Then, the distributed controller (5.3) can be rewritten as

$$\begin{cases} v_i^{\mathrm{od}}(t_k^{T^*})=v_i^{\mathrm{od}}(t_{k-1}^{T^*})+T^*u_k^{v} \\ \omega_i(t_k^{T^*})=\omega_i(t_{k-1}^{T^*})+T^*u_k^{\omega} \end{cases} \tag{5.26}$$

where the vectors $v^{\mathrm{od}}=[v_1^{\mathrm{od}},\cdots,v_N^{\mathrm{od}}]^T$, $\omega=[\omega_1,\cdots,\omega_N]^T$, $u_k^{v}=[u_{1,k}^{v},\cdots,u_{N,k}^{v}]^T$, and $u_k^{\omega}=[u_{1,k}^{\omega},\cdots,u_{N,k}^{\omega}]^T$.

Furthermore, the control inputs (5.16) can be reorganized as

$$\begin{aligned} u_{k+1}^{v}=&-\left(c\tilde{L}^{v}+c\tilde{B}^{v}\odot\Xi^{v}\right)v^{\mathrm{od}}(t_k^{T^*}) \\ &+\left(c\tilde{B}^{v}\odot\Xi^{v}\right)1_N v^{\mathrm{DES}} \end{aligned} \tag{5.27}$$

$$\begin{aligned} u_{k+1}^{\omega}=&-\left(c\tilde{L}^{\omega}+c\tilde{B}^{\omega}\odot\Xi^{\omega}\right)\omega(t_k^{T^*}) \\ &+\left(c\tilde{B}^{\omega}\odot\Xi^{\omega}\right)1_N\omega^{\mathrm{DES}} \end{aligned} \tag{5.28}$$

where the matrices $\tilde{L}^{v}$ and $\tilde{L}^{\omega}$ are respectively defined as following

$$\tilde{l}_{ij}^{v}=\begin{cases} \sum\limits_{q\in\mathcal{N}_i^{v}} a_{iq}^{v}\gamma_{iq}^{v}, & j=i \\ -a_{ij}^{v}\gamma_{ij}^{v}, & j\neq i \end{cases} \tag{5.29}$$

$$\tilde{l}_{ij}^{\omega} = \begin{cases} \sum\limits_{q \in \mathcal{N}_i^{\omega}} a_{iq}^{\omega} \gamma_{iq}^{\omega}, & j = i \\ - a_{ij}^{\omega} \gamma_{ij}^{\omega}, & j \neq i \end{cases} \tag{5.30}$$

It can be obtained by combining (5.26) and (5.27) that

$$\begin{pmatrix} \tilde{v}(t_{k+1}^{T^*}) \\ \tilde{\omega}(t_{k+1}^{T^*}) \end{pmatrix} = \begin{bmatrix} \tilde{H}^{v} & O_{(N+1)\times(N+1)} \\ O_{(N+1)\times(N+1)} & \tilde{H}^{\omega} \end{bmatrix} \begin{pmatrix} \tilde{v}(t_{k}^{T^*}) \\ \tilde{\omega}(t_{k}^{T^*}) \end{pmatrix} \tag{5.31}$$

where $\tilde{v} = [(v^{\text{od}})^T, v^{\text{DES}}]^T$ and $\tilde{\omega} = [\omega^T, \omega^{\text{DES}}]^T$, and the aforementioned matrices are defined as

$$\tilde{H}^{v} = \begin{bmatrix} H^{v} & T^*(cB^{v} \odot \Xi^{v})1_N \\ O_{1\times N} & 1 \end{bmatrix} \tag{5.32}$$

$$\tilde{H}^{\omega} = \begin{bmatrix} H^{\omega} & T^*(cB^{\omega} \odot \Xi^{\omega})1_N \\ O_{1\times N} & 1 \end{bmatrix} \tag{5.33}$$

with $H^{v} = I_N - T^*(c\tilde{L}^{v} + cB^{v} \odot \Xi^{v})$ and $H^{\omega} = I_N - T^*(c\tilde{L}^{\omega} + cB^{\omega} \odot \Xi^{\omega})$.

It is clear that the matrix in (5.31) sanctifies

$$\begin{aligned} &\begin{bmatrix} \tilde{H}^{v} & O_{(N+1)\times(N+1)} \\ O_{(N+1)\times(N+1)} & \tilde{H}^{\omega} \end{bmatrix}^{k} \\ &= \begin{bmatrix} \left(\tilde{H}^{v}\right)^{k} & O_{(N+1)\times(N+1)} \\ O_{(N+1)\times(N+1)} & \left(\tilde{H}^{\omega}\right) \end{bmatrix} \end{aligned} \tag{5.34}$$

If the elements of the control gain matrices γ_{ij}^{v} and γ_{ij}^{ω} satisfy

$$\begin{cases} T^* \left(\sum\limits_{j \in \mathcal{N}_i^{v}} c\bar{a}_{ij}^{v} \gamma_{ij}^{v} + c\bar{a}_{i0}^{v} \gamma_{i0}^{v} \right) < 1, \ \forall i, j \in 1, 2, \cdots, N \\ T^* \left(\sum\limits_{j \in \mathcal{N}_i^{\omega}} c\bar{a}_{ij}^{\omega} \gamma_{ij}^{\omega} + c\bar{a}_{i0}^{\omega} \gamma_{i0}^{\omega} \right) < 1, \ \forall i, j \in 1, 2, \cdots, N \end{cases} \tag{5.35}$$

then, both $\tilde{H}^{v}$ and $\tilde{H}^{\omega}$ have only one eigenvalue $\rho(\tilde{H}^{v}) = \rho(\tilde{H}^{v}) = 1$, which illustrate that each of the digraphs $\hat{G}(A^{v})$ and $\hat{G}(A^{\omega})$ contains a spanning directed tree. Therefore, it can be obtained from the Lemma 5.6 that

$$\begin{aligned}&\lim_{k\to\infty}\begin{bmatrix}\tilde{H}^{v} & O_{(N+1)\times(N+1)}\\ O_{(N+1)\times(N+1)} & \tilde{H}^{\omega}\end{bmatrix}^{k}\\&=\begin{bmatrix}1_{N+1}\left(\beta^{v}\right)^{T} & O_{(N+1)\times(N+1)}\\ O_{(N+1)\times(N+1)} & 1_{N+1}\left(\beta^{\omega}\right)^{T}\end{bmatrix}\end{aligned} \tag{5.36}$$

where $\beta^{v}\geq 0$ and $\beta^{\omega}\geq 0$ are the left eigenvectors of $\tilde{H}^{v}$ and $\tilde{H}^{\omega}$ corresponding to eigenvalue 1, respectively. Thus, they satisfy $1_{N+1}^{T}\beta^{v}=1_{N+1}^{T}\beta^{\omega}=1$.

Based on the form of $(\tilde{H}^{v})^{k}$ and $(\tilde{H}^{\omega})^{k}$, we have

$$1_{N+1}\left(\beta^{v}\right)^{T}=\begin{pmatrix}* & *\\ O_{1\times N} & 1\end{pmatrix} \tag{5.37}$$

$$1_{N+1}\left(\beta^{\omega}\right)^{T}=\begin{pmatrix}* & *\\ O_{1\times N} & 1\end{pmatrix} \tag{5.38}$$

then, we get $\beta^{v}=(0,\cdots,0,1)$ and $\beta^{\omega}=(0,\cdots,0,1)$. Considering (5.31) and (5.36), the voltage regulation and frequency restoration (5.13) can be achieved.

If the elements of the control gain matrices Γ^{v}, Γ^{ω}, Ξ^{v}, and Ξ^{ω} are selected to satisfy the condition (5.35), so that each of the communication digraphs $\hat{G}(A^{v})$ and $\hat{G}(A^{\omega})$ contains a spanning directed tree. Therefore, we can conclude that the voltage amplitude and operating frequency of all DER agents can be synchronized to their desired value under the designed distributed discrete-time voltage and frequency controllers (5.3) with the designed control inputs (5.16).

5.2.2 Active Power Sharing

In this section, we will design the following discrete-time controllers $u_{i,k}^{P}$ by incorporating the relative active power outputs between the neighboring DER agents at terminal time $t_k^{T^*}$ to realize the accurate active power sharing.

$$u_{i,k}^{P}=\sum_{j\in\mathcal{N}_i^{P}}c\gamma_{ij}^{P}a_{ij}^{P}\left[m_j^{P}P_j(t_k^{T^*})-m_i^{P}P_i(t_k^{T^*})\right]/m_i^{P} \tag{5.39}$$

for all $i=1,2,\cdots,N$, where c is a positive constant representing the coupling strength of the cyber network, γ_{ij}^{P} ($i=1,2,\cdots,N$ and $j=0,1,\cdots,N$) is the control gain determining by iteration.

The control objective (5.15) can be achieved asymptotically under the distributed active power controllers (5.39) if their certain conditions are satisfied. Then we will derive the conditions.

Denote the weighted adjacency matrices of active power is $A^P = (a_{ij}^P)_{N\times N} \in [\underline{A}^P, \overline{A}^P]$, and the associated control gain matrices is $\Gamma^P = (\gamma_{ij}^P)_{N\times N}$.

Then, the distributed controller (5.7) can be rewritten as

$$p(t_k^{T^*}) = p(t_{k-1}^{T^*}) + T^* u_k^P \tag{5.40}$$

where $p = (m_1^P P_1, m_2^P P_2, \cdots, m_N^P P_N)^T$ and $u_k^P = (u_{1,k}^P, u_{2,k}^P, \cdots, u_{N,k}^P)^T$.

And the control inputs (5.39) can be rewritten as

$$u_k^P = -c\tilde{L}^P p(t_{k-1}^{T^*}) \tag{5.41}$$

where the matrix $\tilde{L}^P$ is defined as following

$$\tilde{l}_{ij}^P = \begin{cases} \sum\limits_{q\in\mathcal{N}_i^P} a_{iq}^P \gamma_{iq}^P, & j = i \\ -a_{ij}^P \gamma_{ij}^P, & j \neq i \end{cases} \tag{5.42}$$

It can be obtained from (5.40) and (5.41) that

$$p(t_{k+1}^{T^*}) = \left(I_N - T^* c\tilde{L}^P\right) p(t_k^{T^*}) \tag{5.43}$$

If the elements of the control gain matrices γ_{ij}^P satisfies

$$T * \sum_{j\in\mathcal{N}_i^P} c\bar{a}_{ij}^P \gamma_{ij}^P < 1, \ \forall i, j = 1, 2, \cdots, N \tag{5.44}$$

then, the matrix $I_N - T^* c\tilde{L}^P$ has the only one eigenvalue $\rho\left(I_N - T^* c\tilde{L}^P\right) = 1$, which illustrate that the digraph $\hat{G}(A^P)$ contains a spanning directed tree. Therefore, by applying Lemma 4.1, we have

$$\lim_{k\to\infty} \left(I_N - T^* c\tilde{L}^P\right)^k = 1_N \left(\beta^P\right)^T \tag{5.45}$$

considering the normalized left eigenvector $\beta^P \geq 0$ corresponding to eigenvalue 1. Then, it can be deduced from the above denotation of p, that

$$m_1^P P_1 = m_2^P P_2 = \cdots = m_N^P P_N \tag{5.46}$$

Then, the control objective (5.15) can be achieved, if the drop coefficients m_i^P are selected according to the active power ratings of DER$_i$ agent, $P_{i,\max}$.

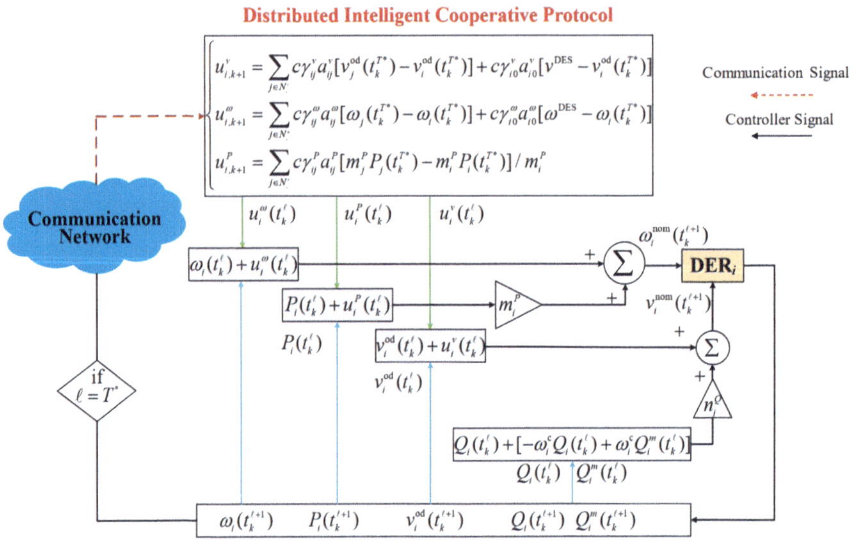

Fig. 5.2 Block diagram of the proposed discrete-time distributed intelligent cooperative protocol

If the elements of the control gain matrices Γ^P is selected to satisfy the condition (5.44), so that the communication digraphs $\hat{G}(A^P)$ contains a spanning directed tree. Therefore, we can conclude that the accurate active power sharing of all DER agents can be achieved under the designed distributed discrete-time active power controllers (5.7) with the designed control inputs (5.39).

Summarize the above, the whole framework of the proposed distributed intelligent cooperative protocol is depicted in Fig. 5.2. It can be observed that the nominal set points v_i^{nom} and ω_i^{nom} can be determined by (5.5) and utilized to synchronize each DER agent's output voltage amplitude v_i^{od} and operating frequency ω_i to their desired reference values v^{DES} and ω^{DES}, respectively.

5.2.3 *Control Implementation*

The detailed implementation of the proposed distributed intelligent cooperative protocol can be represented as follows:

Step1 Initialization: Set the reference trajectory v^{DES} and ω^{DES}, the initial inputs, $u_{i,0}^{v}$, $u_{i,0}^{\omega}$, and $u_{i,0}^{P}$, and the initial states, $v_i^{\text{od}}(0)$, $\omega_i(0)$, and $P_i(0)$ for all $i = 1, 2, \cdots, N$. Let the iteration index $k = 1$ and the tolerance ϵ^{v}, ϵ^{ω} and $\epsilon^{P} > 0$.

Step2 Calculate gain matrices: Determine the gain matrices Γ^{v}, Ξ^{v}, Γ^{ω}, Ξ^{ω}, and Γ^{P}, according to the Algorithm 5.1.

Step3 Calculate nominal set points: Apply $u^{v}_{i,k}$, $u^{\omega}_{i,k}$, and $u^{P}_{i,k}$ to compute $v^{\text{nom}}_{i}(t^{\ell}_{k})$ and $\omega^{\text{nom}}_{i}(t^{\ell}_{k})$, for $\ell = 0, 1, \cdots, T^*$ and $i = 1, 2, \cdots, N$.

Step4 Measure terminal outputs: Apply $v^{\text{nom}}_{i}(t^{T^*}_{k})$ and $\omega^{\text{nom}}_{i}(t^{T^*}_{k})$ to the primary control process and measure the terminal outputs $v^{\text{od}}_{i}(t^{T^*}_{k})$, $\omega_i(t^{T^*}_{k})$, and $P_i(t^{T^*}_{k})$, for $i = 1, 2, \cdots, N$.

Step5 Analyze errors: If $\left|v^{\text{od}}_{i}(t^{T^*}_{k}) - v^{\text{DES}}\right| < \epsilon^{v}$, $\left|\omega_i(t^{T^*}_{k}) - \omega^{\text{DES}}\right| < \epsilon^{\omega}$, and $\left|m^{P}_{i} P_i(t^{T^*}_{k}) - m^{P}_{j} P_j(t^{T^*}_{k})\right| < \epsilon^{P}$ for all $i \neq j = 1, 2, \cdots, N$, then go to step 7; Otherwise go to step 6.

Step6 Calculate control inputs: Let $k = k + 1$, and update $u^{v}_{i,k}$, $u^{\omega}_{i,k}$, and $u^{P}_{i,k}$ according to the protocols (5.16) and (5.39), then go to step 3.

Step7 Stop the iteration.

Algorithm 5.1 Calculate gain matrices Γ^{V}, Ξ^{V}, Γ^{ω}, Γ^{P}

Initialization:

Set $\varepsilon \in (0, 1)$, step length of input update T^*, and let

$\{(\gamma^{V}_{ij})^{N}_{j=0}, (\gamma^{\omega}_{ij})^{N}_{j=0}, \gamma^{P}_{ij}\} \rightarrow \{c(\overline{a^{V}_{ij}})^{N}_{j=0}, c(\overline{a^{\omega}_{ij}})^{N}_{j=0}, c\overline{a^{P}_{ij}}\}$;

Iterative:

while both (12) and (18) hold **do**

$(\gamma^{V}_{ij})^{N}_{j=0} \leftarrow (2-\varepsilon)(\gamma^{V}_{ij})^{N}_{j=0}$, $\gamma^{P}_{ij} \leftarrow (2-\varepsilon)\gamma^{P}_{ij}$, and $(\gamma^{\omega}_{ij})^{N}_{j=0}\} \leftarrow (2-\varepsilon)(\gamma^{\omega}_{ij})^{N}_{j=0}$;

end while

while either (12) or (18) does not hold **do**

$\{(\gamma^{V}_{ij})^{N}_{j=0}, (\gamma^{\omega}_{ij})^{N}_{j=0}, \gamma^{P}_{ij}\} \leftarrow \varepsilon\{(\gamma^{V}_{ij})^{N}_{j=0}, (\gamma^{\omega}_{ij})^{N}_{j=0}, \gamma^{P}_{ij}\}$;

end while

Set $\{\Gamma^{V}, \Gamma^{\omega}, \Gamma^{P}, \Xi^{V}, \Xi^{\omega}\} \rightarrow \{(\gamma^{V}_{ij}), (\gamma^{\omega}_{ij}), (\gamma^{P}_{ij}), (\gamma^{V}_{i0}), (\gamma^{\omega}_{i0})\}$;

By applying an approximate estimation method, the above Step 5 can be implemented in a distributed mode, i.e., each DER agent can guarantee a logic signal which can be utilized to represent DER agent's own error situation, and then send it to its neighbors. Owing to the control inputs designed for each DER agent given in (5.16) and (5.39) will finally converge to zero when the control objectives are achieved, thus they can be used to estimate the errors. And whole error of the control objectives can be reflected in the logic signal for each DER agent after each around of information exchange and iteration update, due to the connectivity of the communication network. Intuitively, the original typologies of the communication digraphs is not corrupted by the selected gain matrices is the key to control objectives' implementation. For example, all the active power gains are always selected to satisfy $\gamma^{P}_{ij} > 0$ when $\mathcal{N}^{P}_{i} \neq \emptyset$ and $j \in \mathcal{N}^{P}_{i}$, otherwise $\gamma^{P}_{ij} = 0$. Also, the voltage and frequency gains γ^{v}_{ij} and γ^{ω}_{ij} should satisfy the same requirements. Thus, all gain matrices can be initialized as their associated communication weights, and further optimize their values by the following Algorithm 5.1.

The coefficient of the gain matrix changing rate $\epsilon \in (0, 1)$ is utilized to make their optimal values to satisfy conditions (5.35) and (5.44). If the initial gain matrices satisfy these conditions, increase them by multiplying $2 - \epsilon > 1$ repeatedly. If the initial gain matrices do not satisfy these conditions, reduce them by multiplying $0 < \epsilon < 1$ repeatedly. This way, the maximum gain matrices to meet the requirements can be calculated by finite iterations. Since the smaller the coefficient ϵ, the less the number of iterations required to calculate the optimal gain matrices, and more optimized values of gain matrices will be calculated with a larger coefficient ϵ, then it is generally necessary to make a tradeoff between the above two factors.

Remark 5.1 In this chapter, the upper bounds of the uncertain communication weights are previously known, thus, the gain matrices only need to calculate once. This way, the proposed distributed intelligent cooperative protocol can be implemented with the same gain matrices throughout the control process. However, if the upper bounds are unknown previously, a distributed control algorithm to obtain the control gain matrices should be designed. This will be our future research direction.

Remark 5.2 As an ideal voltage controller, a tunable compromise between voltage regulation and reactive power sharing should be allowed [193]. Despite, we have designed a continuous time distributed cooperative protocol for voltage regulation, which can not only guarantee the accurate reactive power sharing but also achieve the weighted average voltage of all DER agents to converge to the desired reference value [189]. Unfortunately, it is still difficult to design a tunable voltage and reactive power controller by using discrete time distributed control inputs that are updated in an intermittent communication manner. This will be our future work.

Remark 5.3 The proposed discrete time distributed intelligent cooperative protocol can effectively achieve the perfect tracking of the desired voltage and frequency values. Each DER agent only needs to exchange the information about their owns and some nearest neighbors to achieve synchronization at the end of each round of iteration. Even if there are interval uncertainties of the communication links (caused by the high frequency component of disturbances on voltages or currents) and/or transient power oscillation noises (caused by the flow of large currents), the synchronization can still be achieved if only the network contains a spanning tree. Thus, the proposed distributed control structure can suit interval uncertainty networks, it is more robust against communication disturbances.

Remark 5.4 The gain matrix Γ^V, Ξ^V, Γ^ω, Γ^v and Γ^P are pivotal for ensuring system stability under uncertain communication links. In practice, the initialization of γ_{ij} in Algorithm 5.1 can be set as $\overline{a}_{ij}$, i.e., the lower bounds of the communication weights, to conservatively satisfy conditions (5.35) and ((5.44) from the outset. The coefficient ϵ essentially controls the granularity of the search. A smaller ϵ leads to a coarser search with fewer iterations. Conversely, a larger ϵ yields finer, more optimized gains at the cost of increased computational overhead. The iterative process terminates when the gains are suitable while still respecting the stability inequalities, thus achieving a balance between convergence speed and robustness.

Remark 5.5 While the interval uncertainty model $a_{ij} \in [\overline{a}_{ij}, \bar{a}_{ij}$ effectively captures parametric uncertainties and slow-varying link quality degradation. The, the robustness of our scheme against such non-idealities is intrinsically enhanced by two factors: firstly, the intermittent nature of communication means that the system is tolerant to temporary link failures as long as connectivity is preserved on average over an iteration period; secondly, the discrete-time control law inherently possesses a hold capability, maintaining the last valid control input during communication blackouts. For systems experiencing frequent packet loss, the expectation of the successful transmission rate can be incorporated into the lower bound $\overline{a}_{ij}$ during the design phase.

Remark 5.6 The distributed stopping criterion in Step 5 can be practically implemented by having each DER_i monitor its own state errors. When these errors fall below their respective predefined threshold for a consecutive number of iterations (e.g., 2-3 iterations to avoid premature stopping due to noise), the DER broadcasts a convergence flag to its neighbors. A DER will only cease its iteration permanently once it has itself converged and has received convergence flags from all its neighbors within the communication digraph. This consensus-based stopping mechanism ensures that the entire network agrees on the termination of the control process, preventing isolated nodes from halting prematurely.

5.3 Simulation Results and Discussion

In this section, we evaluate the performance of the proposed discrete time distributed intelligent cooperative control protocol through comprehensive simulations of an islanded microgrid implemented in MATLAB/SimPowerSystems. The test system, represented by the single line diagram in Fig. 5.3, comprises four DER agents,

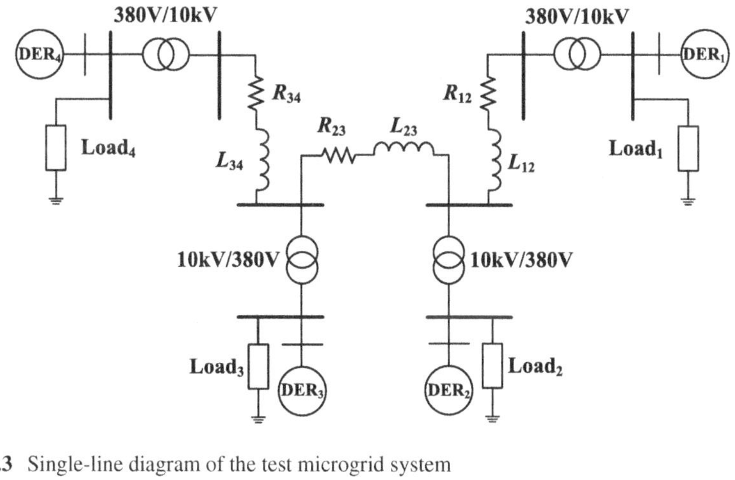

Fig. 5.3 Single-line diagram of the test microgrid system

Table 5.1 The parameters values for microgrid test system

DERs	Parameter	$DER_{1\&3}$ (65 kVA)	Parameter	$DER_{2\&4}$ (115 kVA)
	V^{DC}	800 V	800 V	800 V
	m^P	1×10^{-5}	m^P	0.5×10^{-5}
	n^Q	3×10^{-4}	n^Q	6×10^{-4}
	L^f	2 mH	L^f	2 mH
	C^f	16 μF	C^f	15 μF
	L^o	2 mH	L^o	2 mH
	K^{PV}	10	K^{PV}	8
	K^{IV}	100	K^{IV}	110
	K^{DV}	1.5×10^{-3}	K^{DV}	1.4×10^{-3}
	K^{PC}	5	K^{PC}	4
	K^{IC}	200	K^{IC}	180
Loads	$Load_1$	$Load_2$	$Load_3$	$Load_4$
	P_1 34 kW	P_2 44 kW	P_3 46 kW	P_4 36 kW
	Q_1 34 kVar	Q_2 44 kVar	Q_3 46 kVar	Q_4 36 kVar
Lines	$Line_{1-2}$	$Line_{2-3}$	$Line_{3-4}$	
	R_{12} 0.64 Ω	R_{23} 0.51 Ω	R_{34} 0.58 Ω	
	L_{12} 1.32 mH	L_{23} 1.05 mH	L_{34} 1.21 mH	

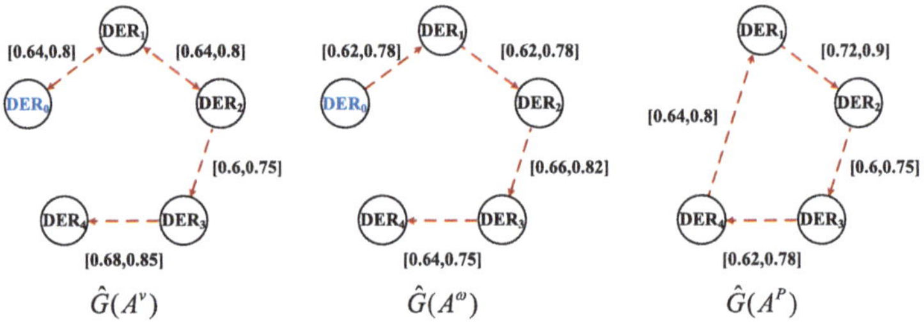

Fig. 5.4 Communication digraphs with interval uncertain weights

four distinct local loads, and three connecting lines. This configuration provides a representative platform for examining the dynamic interactions and control challenges typical of islanded microgrid operations. A complete set of electrical and control parameters for the microgrid test system is systematically detailed in Table 5.1.

The communication topology governing the information exchange among DER agents is characterized by directed graphs with interval uncertain weights, as illustrated in Fig. 5.4. This topology introduces a layer of practical complexity, reflecting the potential uncertainties in real-world communication links. A coupling strength of $c = 1.125$ is selected to ensure adequate interaction dynamics among the agents while maintaining system stability. Notably, the virtual leader agent, denoted as DER_0, is embedded within the digraphs$\hat{G}(A^v)$ and $\hat{G}(A^\omega)$. Its primary role is to generate the desired reference signals to the follower DER agents, specifically

DER_1. Importantly, this leader agent operates in a one way communication manner, which transmits command signals but does not receive any feedback information from the other follower DER agents. The reference values for the microgrid's voltage amplitude and operating frequency are predefined as $v^{\text{DES}} = 380\,\text{V}$ and $\omega^{\text{DES}} = 314\,\text{rad/s}$, respectively. These set points serve as the global control objectives, which the follower DERs must achieve and maintain through local coordination and information sharing in the presence of uncertain communication weights. The forthcoming simulation analysis will demonstrate how effectively the proposed control protocol enables the DER agents to synchronize their outputs to these desired values under the prescribed communication constraints.

The sampling period of the designed discrete-time distributed intelligent cooperative protocol module is set as $T_s = 0.002\,\text{s}$, while the step length of input update is set as $T^* = 100$. As illustrated in Fig. 5.1, this implies that the control input $u^v_{i,k}$, $u^\omega_{i,k}$, and $u^P_{i,k}$ will be updated in every $T_s \times T^* = 0.2\,\text{s}$. Such a configuration strikes a balance between computational efficiency and system responsiveness, ensuring timely updates without overloading the controller. The learning matrices, specifically $\Gamma^v = 0.01A^v$, $\Gamma^\omega = 0.01A^\omega$, $\Gamma^P = 0.01A^P$, $\Xi^v = 0.01B^v$ and $\Xi^\omega = 0.01B^\omega$, which can be calculated by the Algorithm 5.1. In this algorithm, the coefficient for the gain matrix rate of change ϵ is set as 0.9, the total iterative number is 7, and the entire computation process is completed within 0.000344 s. These matrices are robust enough to remain effective even under a reduced coupling strength of $c = 1$, highlighting the adaptability of the proposed method under varying connectivity conditions.

Regarding the communication topologies denoted as $\hat{G}(A^v)$, $\hat{G}(A^\omega)$, and $\hat{G}(A^P)$, the weakest connectivity requirements for each is identical, i.e., the presence of a spanning directed tree. To simplify implementation, it is feasible to unify these three communication digraphs into a single structure with uniform edge weights. Nevertheless, in this study, we intentionally adopt distinct communication digraphs with different weights for each control variable. This design choice aims to rigorously evaluate the robustness of the proposed discrete-time distributed intelligent cooperative protocol when operating over heterogeneous networks with non-identical communication characteristics. To model uncertainties in communication weights, we define a^v_{ij}, as $a^v_{ij} = (1-\alpha)\bar{a}^v_{ij} + \alpha\underline{a}^v_{ij}\text{rand}$. Here, the denotes the degree of uncertainty and and the MATLAB command *rand* generates a random value between 0 and 1, introducing stochastic variations into the communication links. This formulation allows us to assess the control performance under more realistic, non-ideal communication scenarios.

We further investigate the performance of the islanded microgrid test system under four distinct operational scenarios: (1) load changes, (2) plug-and-play capability, (3) link failure, and (4) control parameter test. Each scenario is designed to test specific aspects of system resilience, adaptability, and robustness, thereby offering a comprehensive validation of the proposed control method under both normal and adverse conditions.

5.3.1 Load Change Performance Assessment

In this subsection, we mainly focus on the control performance of proposed discrete-time discrete-time distributed intelligent cooperative protocol in the case of load changes, and the specific simulation process is arranged as follows:

(1) Stage 1: during the time slot of 0−2 s, the microgrid with loads 1−4 is operating in island mode while only the primary control layer is applied.
(2) Stage 2: during the time slot of 2 − 15 s, the microgrid with loads 1 − 4 is operating in island mode while both the primary control and secondary control layers are applied.
(3) Stage 3: during the time slot of 15 − 17 s, the loads 2 and 4 are operating in normal operation and the loads 1 and 3 are removed from the microgrid while only the primary control layer is applied.
(4) Stage 4: during the time slot of 17 − 30 s, the loads 2 and 4 are operating in normal operation and the loads 1 and 3 are removed from the microgrid while both the primary control and secondary control layers are applied.

The simulation results in case of load changes are depicted in Figs. 5.5 and 5.6. It can be observer from Fig. 5.5 that both voltage amplitudes and operating frequencies across all distributed energy resource (DER) agents exhibit deviations from their desired reference values under primary control. These discrepancies arise due to the inherent limitations of decentralized primary control strategies in fully restoring system variables under dynamic conditions. Notably, the active power outputs increase to two distinct levels, which can be attributed to the fact that $\mathrm{DER}_{1,3}$ and $\mathrm{DER}_{2,4}$ share identical control parameters, as specified in Table 5.1. Since frequency constitutes a global variable within the system, it is noteworthy that all DER agents ultimately synchronize to a common frequency value, despite transient mismatches.

A significant improvement occurs when the proposed discrete-time distributed intelligent cooperative control protocol is activated at at $t = 2$ s. As depicted in Fig. 5.5a, b, the voltage and frequency deviations are rapidly eliminated. This demonstrates the effectiveness of the secondary control layer in achieving secondary control objectives. Simultaneously, the active power sharing profile established during the primary control phase is preserved, as evidenced in Fig. 5.5c. This indicates that the active power sharing can still successfully be achieved under the proposed method.

During the stage 3 of the experiment, local load_1 and load_3 are disconnected from the microgrid system at $t = 15$ s. This disturbance leads to an immediate increase in the voltage amplitudes of $\mathrm{DER}_{1,3}$ agents, resulting from the sudden reduction in local power demand. Due to coupled power flow dynamics, the voltage amplitudes of $\mathrm{DER}_{2,4}$ agents experience a momentary dip at the same time instant. These interactions highlight the interlinked nature of networked microgrid systems. When the proposed discrete-time distributed intelligent cooperative protocol is reactivated at $t = 17$ s, all DER agents promptly recover their voltage and frequency set points, while maintaining accurate active power sharing (as shown in Fig. 5.5) according to

Fig. 5.5 State evolution for the proposed discrete-time distributed intelligent cooperative protocol in case of load change performance assessment: (**a**) voltage amplitude; (**b**) operating frequency; (**c**) active power output (©2024 IEEE. Reprinted, with permission, from IEEE Transactions on Industrial Informatics [80])

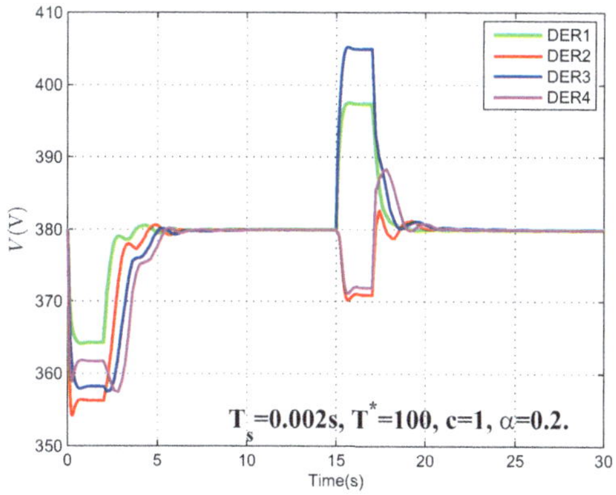

(a)

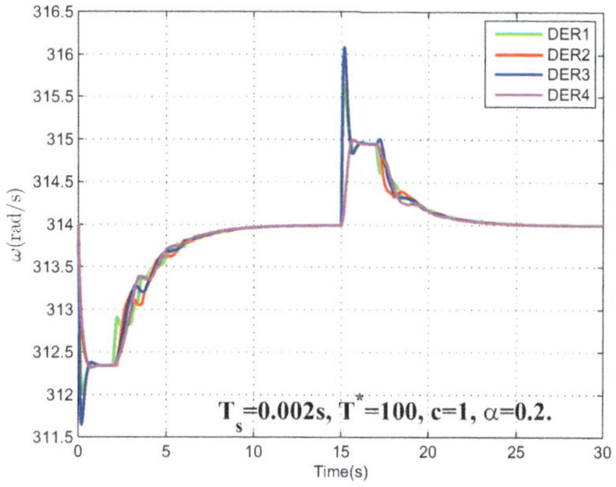

(b)

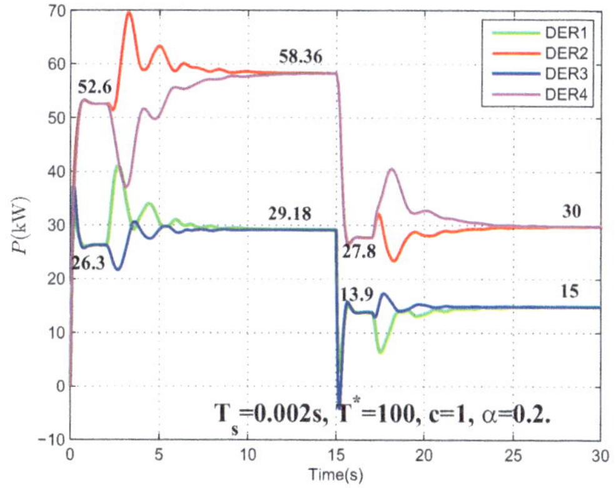

(c)

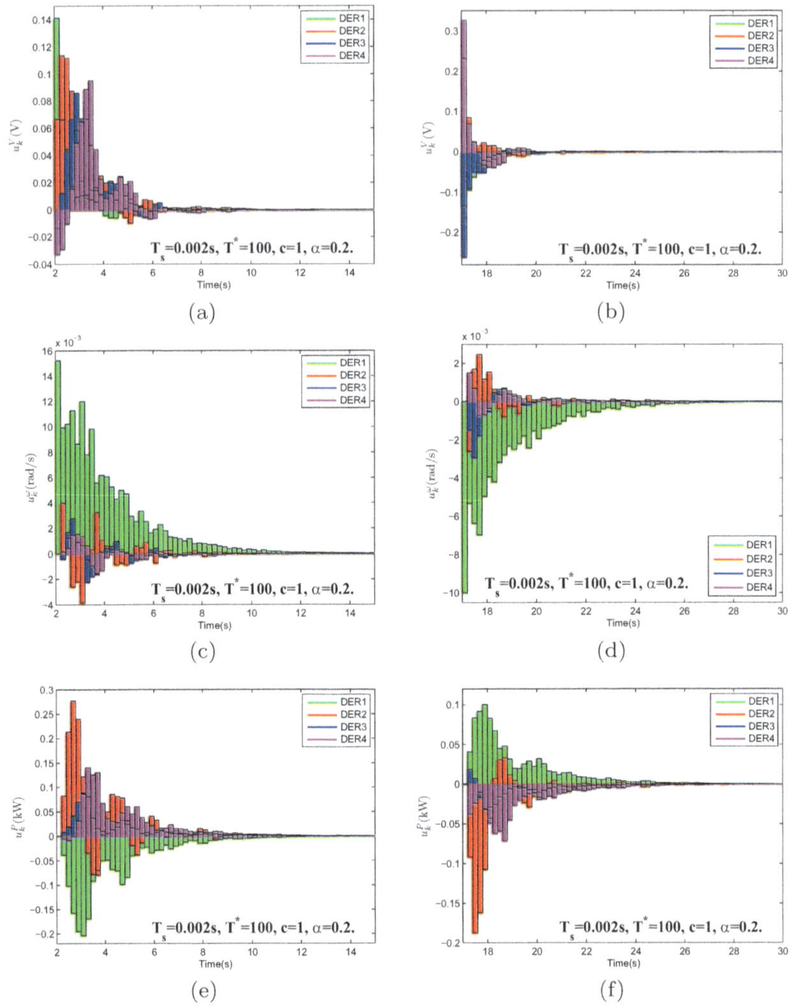

Fig. 5.6 Control input evolution for the proposed discrete-time distributed intelligent cooperative protocol in case of load change performance assessment: (**a**) voltage input; (**b**) voltage input; (**c**) frequency input; (**d**) frequency input; (**e**) active power input; (**f**) active power input (©2024 IEEE. Reprinted, with permission, from IEEE Transactions on Industrial Informatics [80])

their respective droop coefficients. This consistent performance under varying load conditions demonstrates the robustness of the proposed strategy.

The evolution of the distributed control inputs of all DER agents is presented in Fig. 5.6. Among the DER agents, the DER_1 exhibits the fastest dynamic response and the largest control effort amplitude. This is because DER_1 is directly connected to the virtual leader agent DER_0, which serves as the reference source for the entire system. With the specified input update period of $T_s \times T^* = 0.2$ s, the control inputs are updated totally 65 times during each of the two operational intervals: from $2\,\text{s} \le t \le 15\,\text{s}$ and $17\,\text{s} \le t \le 30\,\text{s}$. Once voltage regulation and frequency restoration are achieved, the control inputs converge to zero. It is also notable that the control signals exhibit non-smooth and stochastic variations during transients. These characteristics stem from the presence of uncertain communication links.

Overall, the simulation results validate that the proposed discrete-time distributed cooperative protocol not only restores voltage and frequency under significant load changes but also maintains accurate power sharing between all DERs in the presence of realistic communication imperfections.

5.3.2 *Plug-and-Play Capability*

In this subsection, we mainly focus on the control performance of proposed discrete-time distributed intelligent cooperative protocol in the case of DER plug-and-play. the microgrid with loads 1–4 is operating in island mode while both the primary control and secondary control layers are applied. The specific simulation process is arranged as follows:

(1) At $t = 15\,\text{s}$, DER_4 agent is plugged out from the microgrid.
(2) At $t = 24\,\text{s}$, DER_4 agent is plugged back to the microgrid.

The simulation results demonstrating the plug-and-play capability of the proposed control method are illustrated in Figs. 5.7 and 5.8. It can be seen from Eq. 5.16 that the active power output of DER_4 suddenly decreases to 0 upon disconnection. Despite this sudden change, both its voltage amplitude and operating frequency exhibit a brief transient variation before successfully recovering to their desired values. This recovery demonstrates the robustness of the control strategy in the presence of DER's plug-and-play. Figure 5.8 further reveals that the control inputs of the DER_4 agent remain constant during the interval $t \in [15, 24]\,\text{s}$. This stagnation occurs because the communication links associated with coefficients a_{43}^{v}, a_{43}^{ω}, a_{43}^{P}, and a_{14}^{P} are disrupted. Nevertheless, the rest of the system comprising $DERs_{1-3}$ maintains operational integrity. Thanks to the remaining communication topology, as depicted in Fig. 5.4, which still satisfies the necessary connectivity conditions, the voltage amplitudes and operating frequencies of $DERs_{1-3}$ can be effectively regulated back to their reference values.

Moreover, a new distribution pattern of active power sharing is established among the remaining three $DERs_{1-3}$, as depicted in Fig. 5.7c. This redistribution of active power is achieved through the discrete time distributed intelligent cooperative protocol, which ensures proportional load sharing according to each DER's droop coefficients. When the DER_4 agent is plug back to the microgrid at $t = 24\,\text{s}$, it promptly resumes its original power output level. Consequently, the initial active power sharing pattern among all four DER agents is fully restored, as shown in Fig. 5.7c.

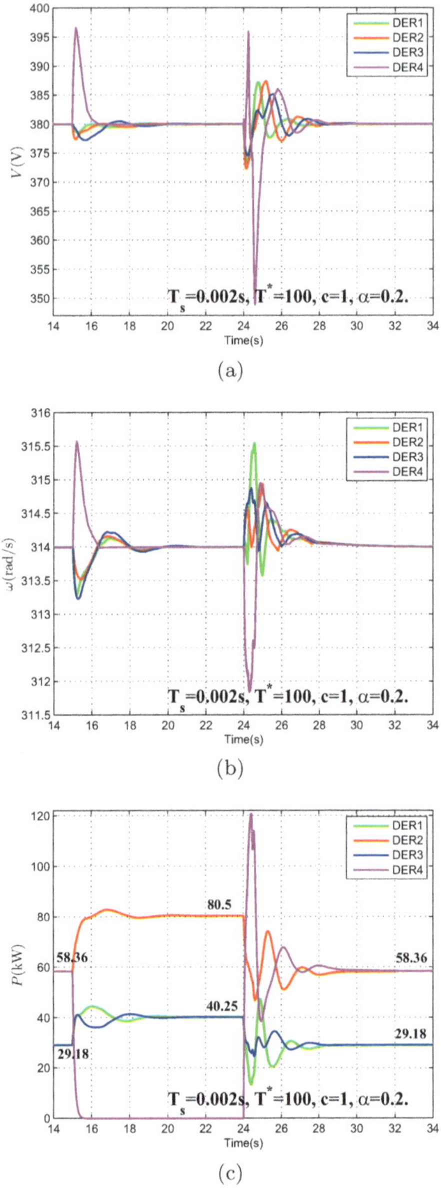

Fig. 5.7 Control input evolution for the proposed discrete-time distributed intelligent cooperative protocol in case of plug-and-play capability: (**a**) voltage amplitude; (**b**) operating frequency; (**c**) active power output (©2024 IEEE. Reprinted, with permission, from IEEE Transactions on Industrial Informatics [80])

Notably, the voltage amplitudes and operating frequencies across all agents experience only minor and short-lived deviations before quickly returning to their desired values. After $t = 30$ s, the entire microgrid system exhibits excellent performance in voltage regulation, frequency restoration, and accurate active power sharing. These results confirm that the proposed discrete time distributed control

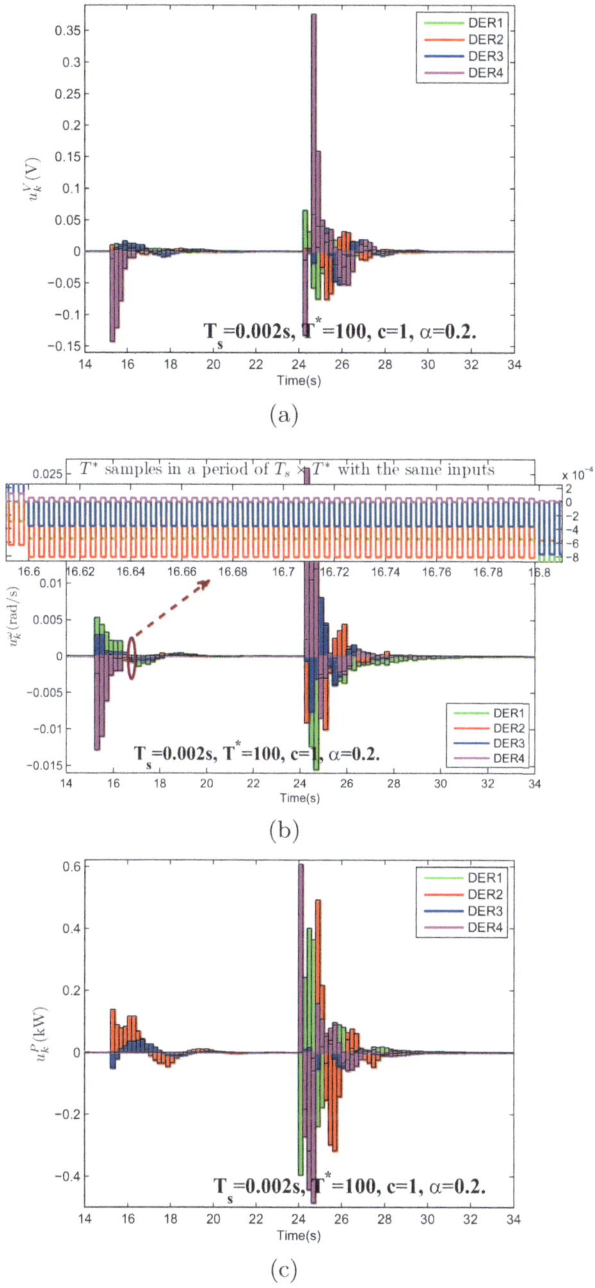

Fig. 5.8 Control input evolution for the proposed discrete-time distributed intelligent cooperative protocol in case of plug-and-play capability: (**a**) voltage input; (**b**) frequency input; (**c**) active power input (©2024 IEEE. Reprinted, with permission, from IEEE Transactions on Industrial Informatics [80])

protocol is not only effective under normal conditions but also highly adaptive to communication topological changes, thereby validating its plug-and-play capability.

It is important to note that each DER inverter is equipped with a self-protection mechanism, such as saturation constraints, to limit its power output and prevent overloading. When a DER agent reaches its maximum allowable power, it switches to a constant power operating mode and ceases to participate in the distributed communication network. The remaining load is then automatically redistributed among the other DERs, leading to a new power sharing equilibrium. To illustrate this feature, the total load is increased by 4 kW + j4 kVar, and then, the DER_4 is plug out from the microgrid at $t = 15$ s. According to the original allocation pattern of power sharing, the active power output of DER_2 agent will beyond it's capacity. As a result, the DER_2 agent will switch to the special operation mode and keep its constant active power output as it achieves its limit 81.3 kW as shown in Fig. 5.9c. Then, the remaining power demand is then automatically reallocated between the rest DER_1 and DER_3 agents.

Similarly, a voltage regulator with reactive power constraints should be adopted to limit the reactive power output and prevent it from being overloaded. As depicted

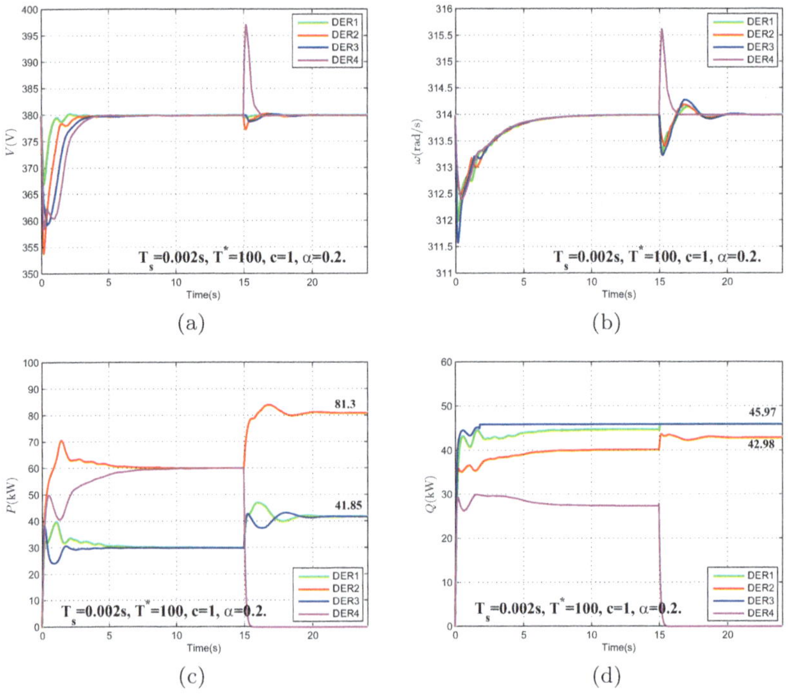

Fig. 5.9 State evolution for the proposed discrete-time distributed intelligent cooperative protocol in case of the power outputs of some DER agents achieve their limits: (**a**) voltage amplitude; (**b**) operating frequency; (**c**) active power output; (**d**) reactive power output (©2024 IEEE. Reprinted, with permission, from IEEE Transactions on Industrial Informatics [80])

in Fig. 5.9d, the reactive power outputs of DER_2 and DER_4 reach their respective limits of 45.97 kVar. Despite these saturations, the system maintains excellent voltage regulation and frequency restoration performance, as evidenced in Fig. 5.9a. This further attests to the resilience and flexibility of the proposed control method under various operational constraints.

5.3.3 Link Failure Test

In this subsection, we conduct a comparative analysis to evaluate the robustness of the proposed discrete time distributed intelligent cooperative protocol against communication link failures by comparing its performance against the conventional centralized secondary control scheme as referenced in [135, 191–219] under the same discrete time control framework. Both control strategies are implemented and tested within the same discrete-time control framework to ensure a fair comparison. While it is known that the vulnerability of centralized control architectures with respect to a controller failure can be coped using model predictive control techniques [229]. Furthermore, we only test the link failure situation here. For clarity and consistency, we revisit the load increase scenario previously introduced in Sect. 5.3.1. In this setup, the microgrid supplying loads 1–4 operates in islanded mode with primary control, and secondary control layers activated at $t = 2$ s.

The specific simulation process is arranged as follows:

(1) At $t = 3$ s, the link between DER_1 and DER_0 agents is broken.
(2) At $t = 9$ s, the link between DER_1 and DER_0 agents is reconnected.
(3) At $t = 10$ s, the link between DER_2 and DER_3 agents is broken.
(4) At $t = 13$ s, the link between DER_2 and DER_3 agents is reconnected.

Under the centralized secondary control framework, a single controller (associated with the DER_0 agent) has access to real time measurements of voltage amplitudes, operating frequencies, and active power outputs from all DER agents. To simulate realistic communication failures, it can be assumed that measurements from the DER_1 agent become unavailable to the central controller during the interval $3\,\text{s} < t < 9\,\text{s}$. Subsequently, data from the DER_3 agent are also lost during the time stage $10\,\text{s} < t < 13\,\text{s}$. These interruptions mimic partial communication breakdowns that may occur in practical microgrid operations.

The simulation results in case of link failure test are illustrated in Figs. 5.10 and 5.11. From the results, it is evident that both the discrete time distributed intelligent cooperative protocol and centralized secondary control scheme experience a slowdown in the restoration of voltage, frequency, and active power outputs whenever a communication link is broken. Once the link is restored, the system resumes normal regulation behavior. Notably, as shown in Fig. 5.11a, c and e, the centralized control scheme exhibits faster transient recovery compared to the distributed approach depicted in Fig. 5.10a, c and e. This performance advantage can be attributed to the high bandwidth of communication links typically employed

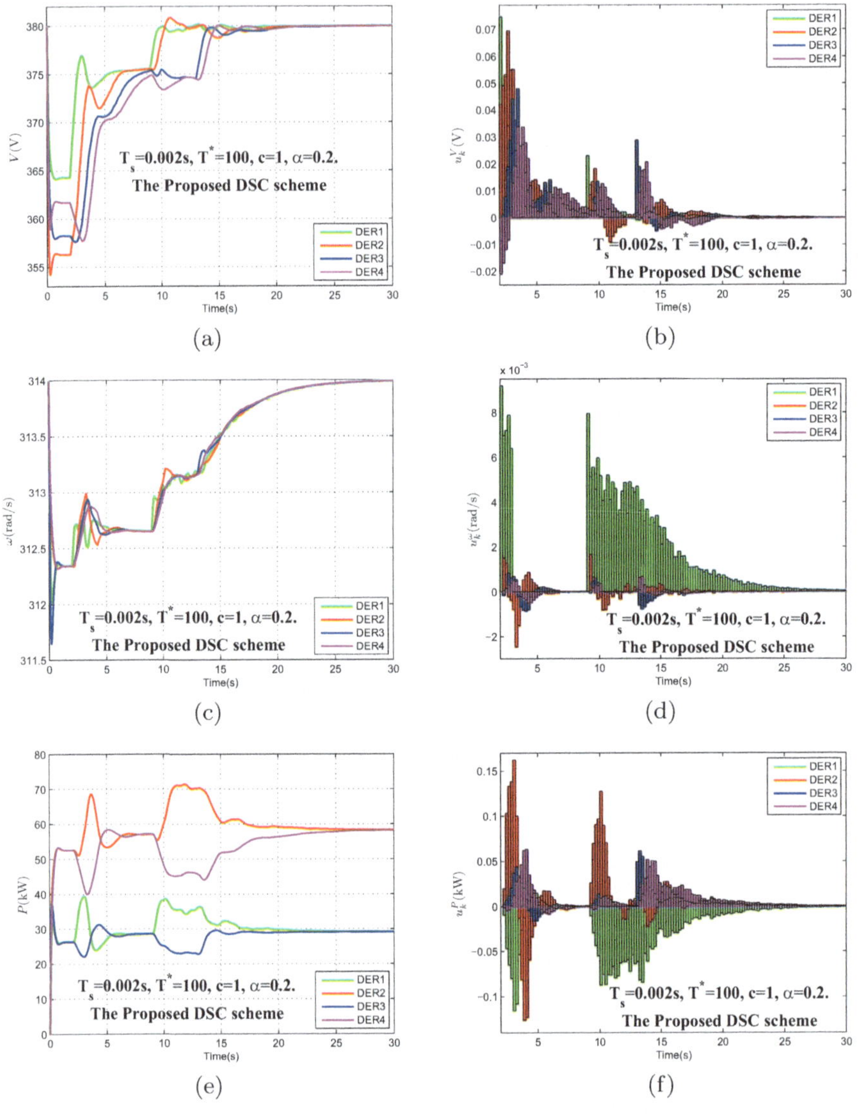

Fig. 5.10 State and control input evolution for the proposed discrete-time distributed intelligent cooperative protocol in case of link failure: (**a**) voltage amplitude; (**b**) voltage input; (**c**) operating frequency; (**d**) frequency input; (**e**) active power output; (**f**) active power input (©2024 IEEE. Reprinted, with permission, from IEEE Transactions on Industrial Informatics [80])

in centralized systems, which allows quicker data aggregation and control signal dissemination. However, a deeper examination of the control input signalsspecifically, Fig. 5.10b, d and f for the distributed protocol and Fig. 5.11b, d and f for the centralized one, reveals a significant drawback of the centralized scheme: it leads to

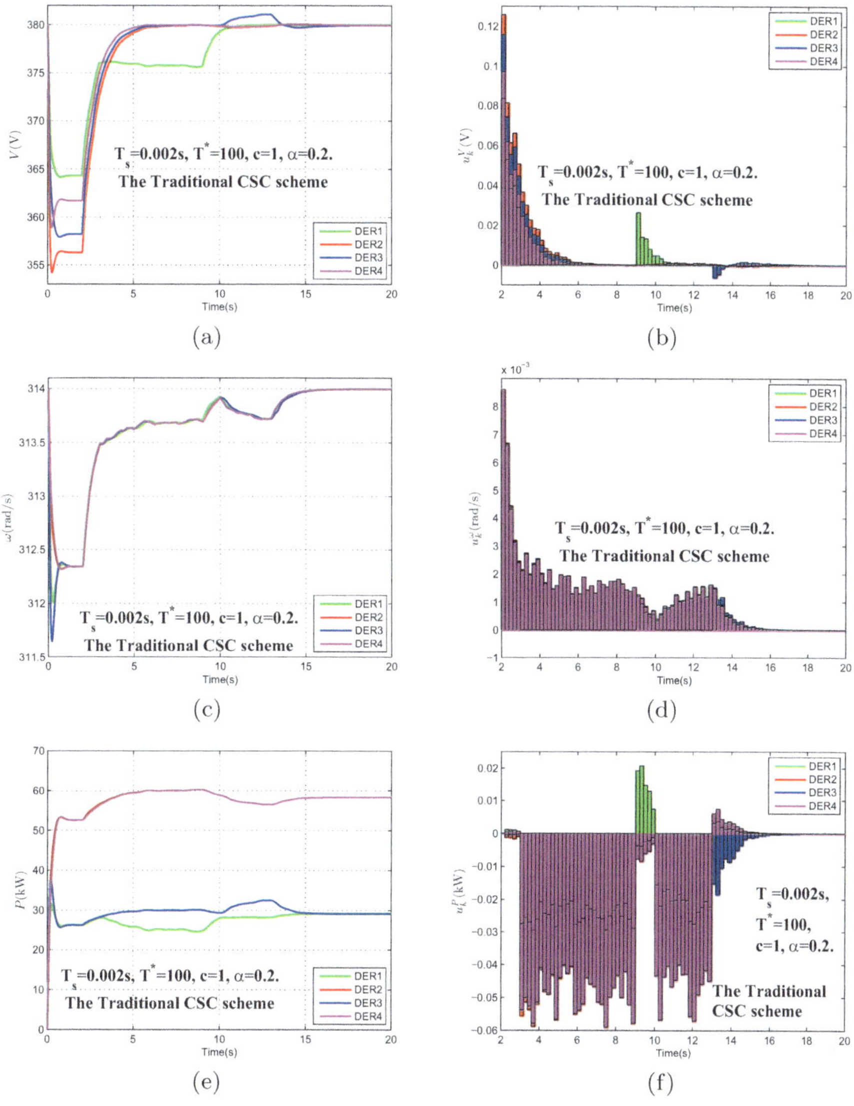

Fig. 5.11 State and control input evolution for the centralized secondary control scheme in case of link failure: (**a**) voltage amplitude; (**b**) voltage input; (**c**) operating frequency; (**d**) frequency input; (**e**) active power output; (**f**) active power input (©2024 IEEE. Reprinted, with permission, from IEEE Transactions on Industrial Informatics [80])

more pronounced imbalances among the control inputs of individual DER agents. Such disparities may induce imbalance on converters and reduce the overall system resilience. In contrast, the distributed protocol facilitates a more balanced sharing of control effort, thereby promoting equitable participation among all DERs. Although both control architectures possess distinct merits suited to different application contexts, the proposed discrete time distributed intelligent cooperative protocol demonstrates particular advantages in microgrids comprising a large number of DER agents. Its implementation relies on a sparse communication network, which not only distributes computational and communication loads evenly across agents but also alleviates the risk of single-point failures. This feature becomes increasingly critical as the system scales up, ensuring both operational reliability and economic feasibility in modern decentralized energy systems.

5.3.4 Control Parameter Test

In order to clarify the effect of the control parameters including the sampling period T_s, the step length of input update T^*, the coupling strength c, and and the degree of uncertainty α on the global control performance of the microgrid system, the microgrid test system with different control parameters will be simulated in this section. For simplicity, only the load increasing scenario in Sect. 5.3.1 is considered. The simulation results in case of control parameter test are depicted in Figs. 5.12, 5.13, 5.14, 5.15, 5.16, and 5.17.

The state evolution of the microgrid test system under the proposed discrete time distributed intelligent cooperative protocol is investigated under various parameter settings, including different sampling periods different sampling period T_c, step length of input update T^* and coupling strength c. The corresponding simulation results are respectively depicted in Eqs. 5.34–5.36. A comparative analysis of the simulation results reveals that an increase in coupling strength significantly enhances the convergence rate. For instance, when the coupling strength is set to $c = 1.125$, the system exhibits markedly faster convergence compared to scenarios with lower coupling strengths, as clearly observed in Fig. 5.12a, c, e and b, d, f. These results illustrates the critical role of coupling strength in accelerating system dynamics.

However, the performance of the proposed protocol is highly sensitive to the step length of the input update. When T^* is increased to 300, the larger step length of input update leads to the failure of the proposed discrete time distributed intelligent cooperative protocol. The voltage regulation, frequency restoration and active power sharing could not be realized any more as shown in Fig. 5.14, due to the system is unable to meet the necessary conditions specified in Eqs. (5.35) and (5.44), leading to a breakdown in cooperative control. This highlights the importance of selecting an appropriately input update period T^*. In fact, varying input update periods $T_s \times T^*$ (e.g., 0.1 s in Fig. 5.12, 0.2 s in Fig. 5.5, and 0.6 s in Fig. 5.14) result in distinctly different control performances. To approximate the satisfaction of stability conditions (5.35) and (5.44) under extreme operating conditions, the sampling period T_s must be chosen sufficiently. Nevertheless, in practical applications, a careful trade off between control costs and system performance is essential.

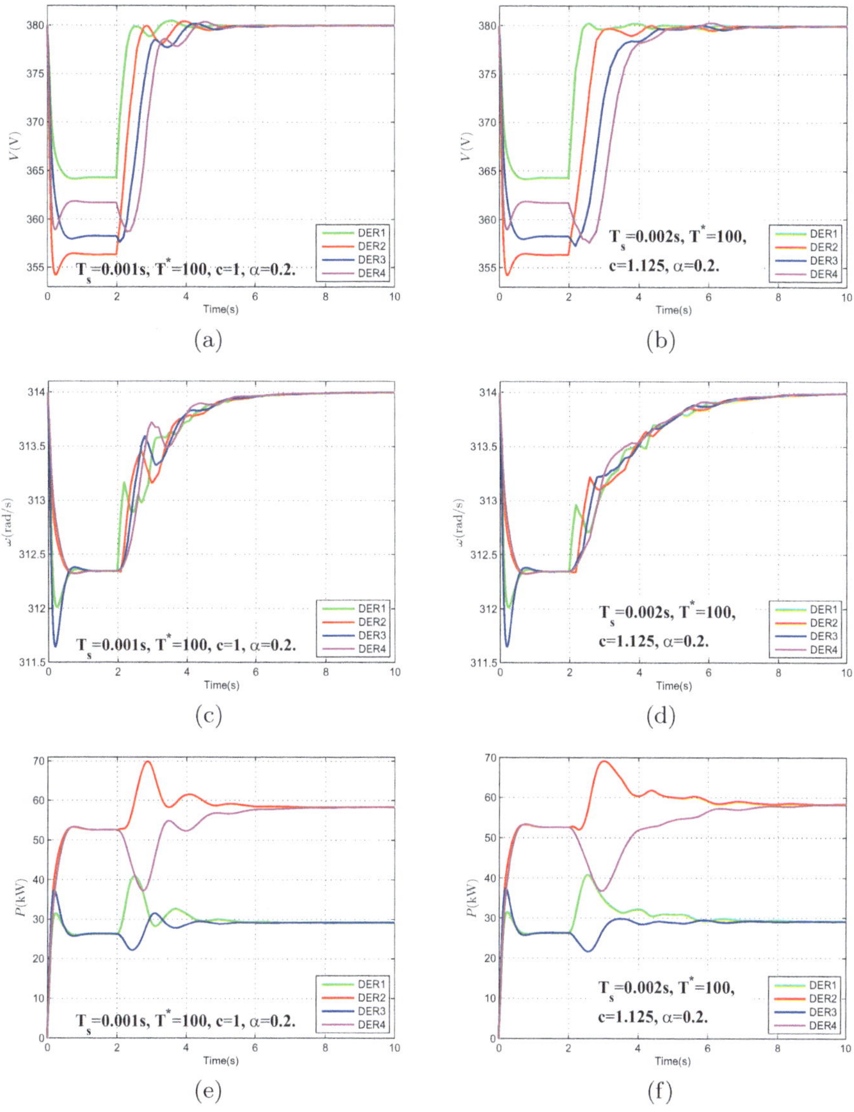

Fig. 5.12 State evolution for the proposed discrete-time distributed intelligent cooperative protocol with different T_s and c: (**a**) voltage amplitude; (**b**) voltage amplitude; (**c**) operating frequency; (**d**) operating frequency; (**e**) active power output; (**f**) active power output (©2024 IEEE. Reprinted, with permission, from IEEE Transactions on Industrial Informatics [80])

Further simulations examine the impact of communication link reliability on the microgrid consensus behavior. Figure 5.15 compares the state evolution under a fixed communication link $\alpha = 0$ and an uncertain communication link $\alpha = 0.3$. The results indicate that a lower degree of uncertainty corresponds to a faster

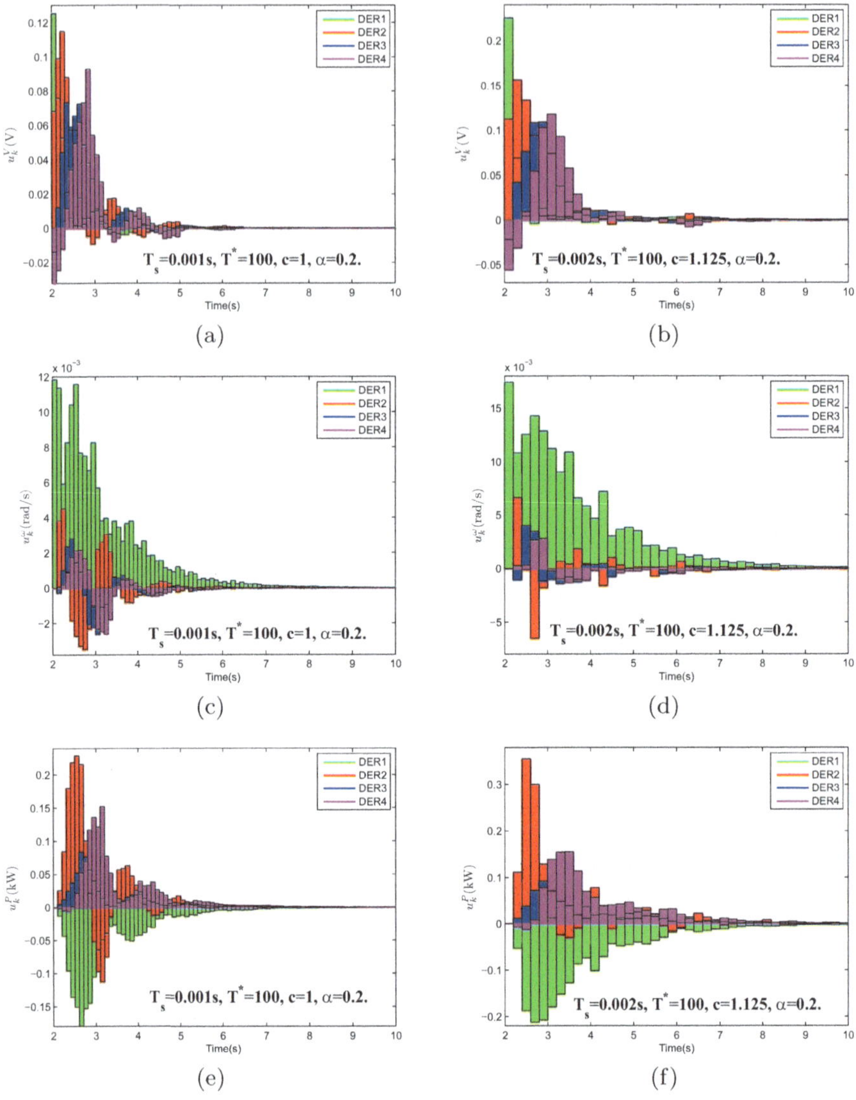

Fig. 5.13 Control input evolution for the proposed discrete-time distributed intelligent cooperative protocol with different T_s and c: (**a**) voltage input; (**b**) voltage input; (**c**) frequency input; (**d**) frequency input; (**e**) active power input; (**f**) active power input (©2024 IEEE. Reprinted, with permission, from IEEE Transactions on Industrial Informatics [80])

convergence speed, as corroborated by the outcomes in Fig. 5.15. Moreover, as shown in Fig. 5.16, control inputs are updated more smoothly when the uncertainty level is reduced, contributing to improved transient performance.

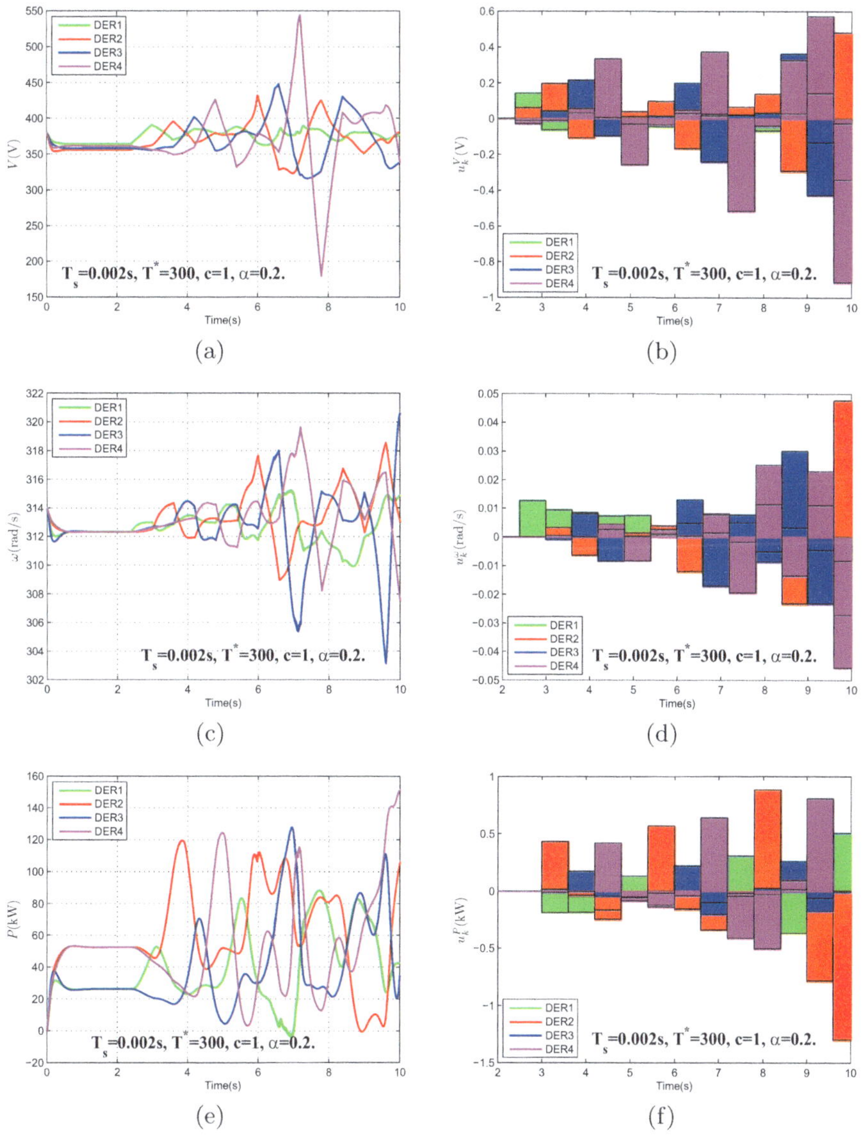

Fig. 5.14 State and control input evolution for the proposed discrete-time distributed intelligent cooperative protocol with a larger T^*: (**a**) voltage amplitude; (**b**) voltage input; (**c**) operating frequency; (**d**) frequency input; (**e**) active power output; (**f**) active power input (©2024 IEEE. Reprinted, with permission, from IEEE Transactions on Industrial Informatics [80])

To evaluate the robustness of the proposed discrete-time distributed intelligent cooperative protocol under randomly switching communication links, additional tests are conducted with stochastically varying communication weights. For exam-

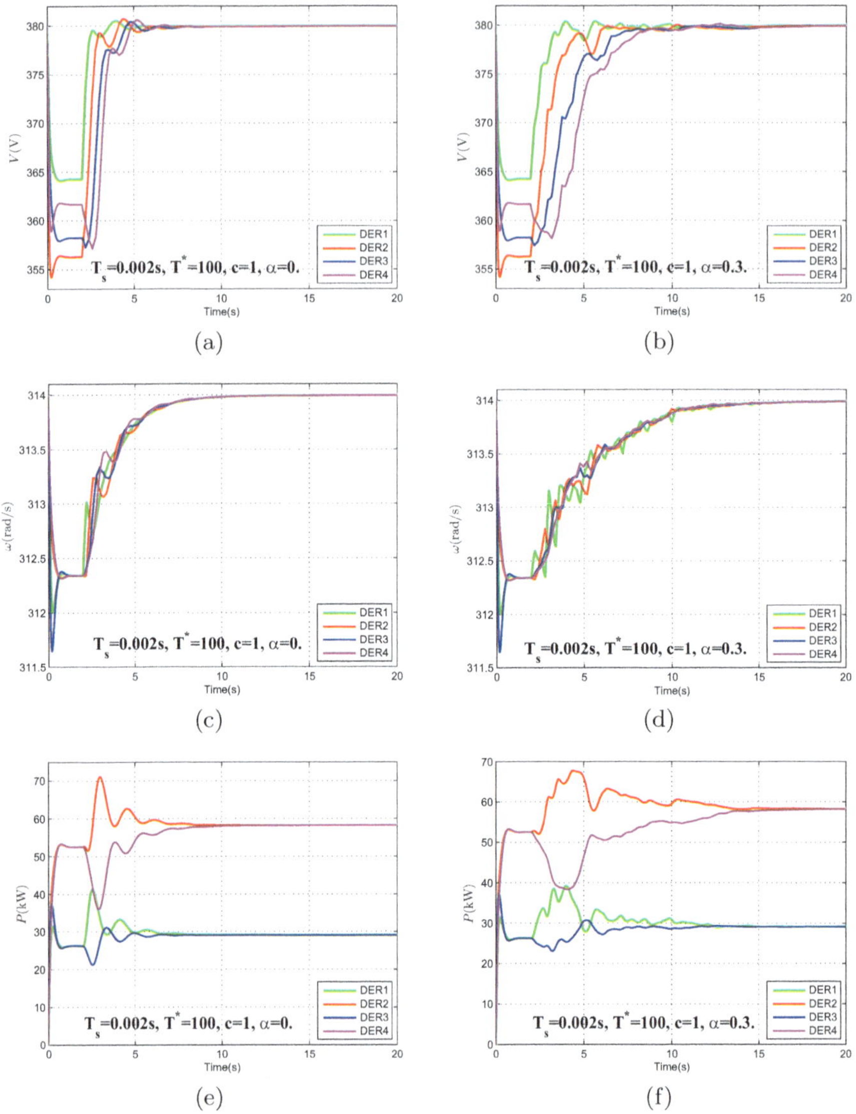

Fig. 5.15 State evolution for the proposed discrete-time distributed intelligent cooperative protocol with different α: (**a**) voltage amplitude; (**b**) voltage amplitude; (**c**) operating frequency; (**d**) operating frequency; (**e**) active power output; (**f**) active power output (©2024 IEEE. Reprinted, with permission, from IEEE Transactions on Industrial Informatics [80])

ple, the weight a_{ij}^v is modeled as $a_{ij}^v = \bar{a}_{ij}^v$randint, where the MATLAB command *randint* can randomly generate an element either 0 or 1. Under such conditions, the connectivity of the communication network may not guaranteed. Despite this, the simulation results in Fig. 5.17 demonstrate that the microgrid test system under

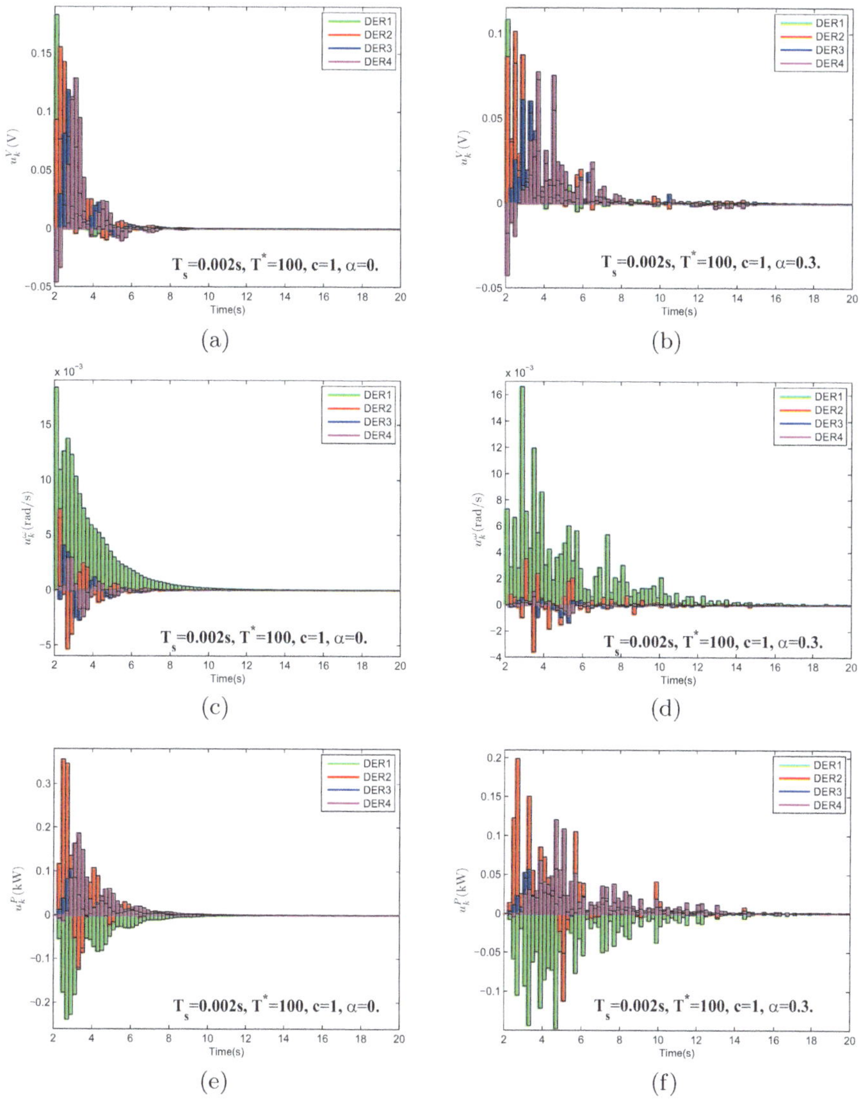

Fig. 5.16 Control input evolution for the proposed discrete-time distributed intelligent cooperative protocol with different α: (**a**) voltage input; (**b**) voltage input; (**c**) frequency input; (**d**) frequency input; (**e**) active power input; (**f**) active power input (©2024 IEEE. Reprinted, with permission, from IEEE Transactions on Industrial Informatics [80])

the proposed discrete-time distributed intelligent cooperative protocol maintains excellent voltage regulation, frequency restoration, and active power sharing. This robustness can be attributed to the fact that the switching frequency of the random communication links is governed by the sampling period T_s. Therefore, by

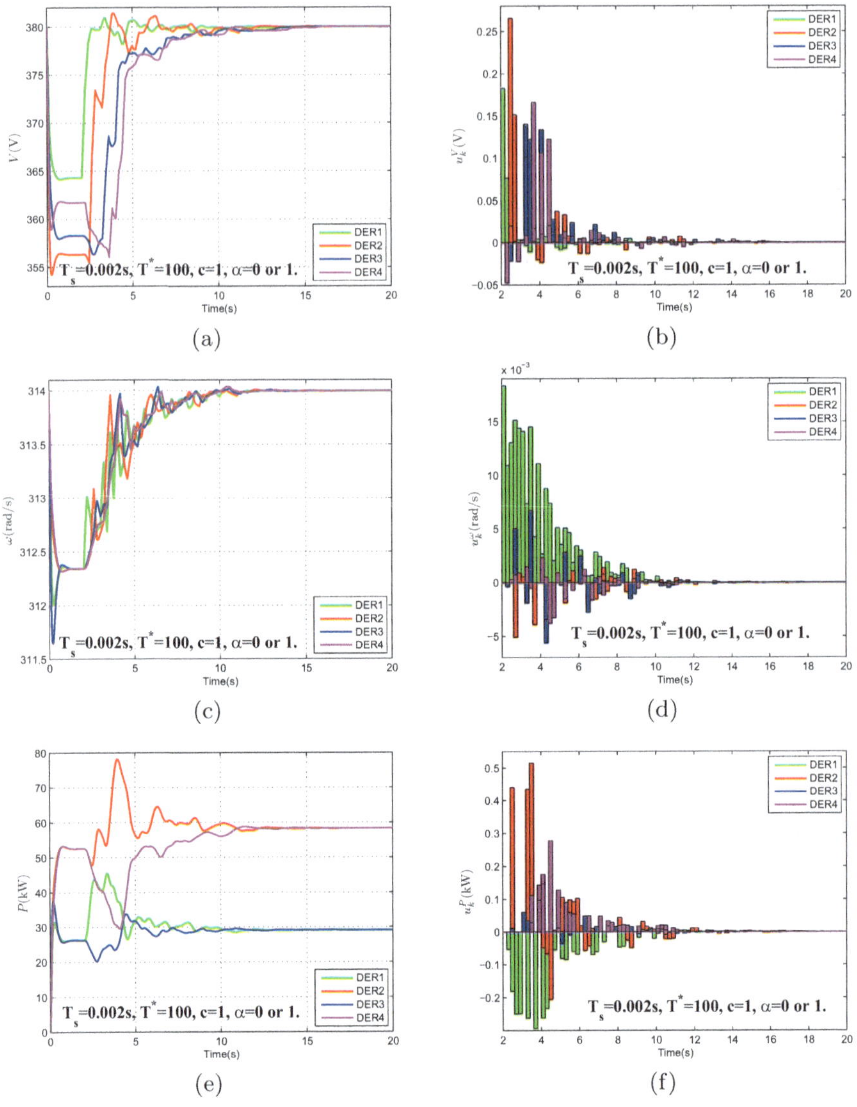

Fig. 5.17 State and control input evolution for the proposed discrete-time distributed intelligent cooperative protocol with $\alpha = 0$ or 1: (**a**) voltage amplitude; (**b**) voltage input; (**c**) operating frequency; (**d**) frequency input; (**e**) active power output; (**f**) active power input (©2024 IEEE. Reprinted, with permission, from IEEE Transactions on Industrial Informatics [80])

appropriately specifying the sampling period, the proposed discrete time distributed intelligent cooperative protocol remains effective even in sparse networks with randomly switching topologies. Once again, a balanced consideration of control costs and performance is advisable in practical implementations.

Remark 5.7 The proposed control aims to achieve the voltage regulation, i.e., voltage amplitudes of all DER agents must be synchronized to their desired values. However, in some extreme operating conditions, all DER agents' voltages must be adjusted to prevent the circuits that make up the microgrid from being overloaded. In this situation, the proposed discrete-time distributed intelligent cooperative protocol must be extended with the capability to handle excessive currents in microgrid branches. In detail, the current errors among the branches will be injected into the local voltage controllers by the current information exchanges among DER agents. This will be our future work.

Remark 5.8 The proposed discrete-time distributed intelligent cooperative protocol can also be extended to control voltages at buses with loads but without DER agents. Firstly, the microgrid model with heterogeneous nodes, including DER agents, loads, and energy storage systems (ESSs), and so on can be reestablished as a new network. And the corresponding discrete-time distributed intelligent cooperative protocol may be multistage, hybrid, and even heterogeneous due to the different dynamics and time scales of each node possesses. Alternatively, every node (i.e., DER agent, load bus, ESS) in the microgrid can be assigned with an agent, which can automatically manage its own operation mode and exchange its measurements with neighbors. Finally, the proposed discrete-time distributed intelligent cooperative protocol can also be extended to control voltages at buses with loads without DER agent by imposing different action properties on these heterogeneous agents.

5.4 Summary

A robust discrete-time distributed intelligent cooperative protocol has been proposed for the voltage regulation and frequency restoration as well as accurate active power sharing of an AC microgrid on the condition of directed communication links with interval uncertainties. In the proposed distributed control structure, all DER agents in an microgrid only need to access limited information of the system parameters, and communicate with their neighbors intermittently. Sufficient conditions containing the digraph connectivity requirements and learning gain inequalities that guarantee the system stability have been derived. This way, the robust capability of the AC microgrid to both the internal uncertainties and the external disturbances can be guaranteed under the proposed discrete-time distributed intelligent cooperative protocol. Finally, simulation results coincide with the present control methods.

Chapter 6
Resilient Distributed Cooperative Intelligence for AC Microgrids Subject to Disturbances

This chapter proposes a resilient distributed multi-agent control scheme for AC microgrids subject to additive noise and time-delay disturbances. The proposed multi-agent control scheme is composed of three distributed consensus protocols, which is able to synchronize the output voltages and frequencies of inverter-based DERs to their reference values and achieve the optimal active power-sharing property by a low bandwidth communication network with noise and time-delay disturbances in almost sure convergence. By means of the stochastic analysis tools and algebraic graph theory, distributed consensus control protocols are designed to be employed for the secondary control layer of microgrids. On this basis, we deduce the stability criteria of the closed-loop microgrid system under noise and time-delay disturbances. As a result, the proposed consensus protocols can well restore the voltage and frequency derivation produced at the primary control layer, meanwhile, can well achieve the optimal power sharing even though there exist communication disturbances. Several simulation scenarios on an islanded microgrid are provided to verify the control performance.

The microgrid composed of multiple intelligent nodes is a typical smart power network system that can achieve autonomous control and communication behaviors to provide power and electricity to the demand side [112, 230, 231]. There are two operating modes of microgrids are clearly shown, namely, islanding operation under adverse conditions and grid connected operation when the point of common coupling (PCC) is off/on connection, respectively [232, 233].

The hierarchical control framework that adopts three-layer control has provided a globally stable paradigm for microgrids in recent years [110, 234], including tertiary control [235], secondary control [236], and primary control [237]. In the droop-based primary control scheme, automatic voltage and frequency regulation is adopted to achieve voltage and frequency stability, and the sharing of active and reactive power is the main design concept [238, 239]. However, voltage and frequency regulation methods based on droop control will produce significant deviations, which requires accurate regulation by secondary control [240, 241].

J. Lai, X. Lu, *Distributed Cooperative Intelligence in Microgrid Control*,
https://doi.org/10.1007/978-981-95-6884-0_6

And in the secondary control strategy, decentralized [109] and distributed control [109, 120, 132, 234, 242–244] are the main two implementation schemes.

Distributed multi-agent technology has shown outstanding advantages in communication costs, robustness of controller single point of failure, future expansion of microgrids, and implementation of plug-and-play [245]. Therefore, it is widely used for solving microgrid problems in recent works [246, 247]. Specifically, all distributed energy resource agents can achieve distributed control objectives by interacting with neighbors in a low bandwidth network. The tertiary control focuses on power flow control and optimization scheduling problems [248].

This chapter will introduce a distributed secondary control method for islanded microgrids, where the secondary control is absorbed in compensating for steady-state errors in frequency and voltage generated by the primary control layer [109, 240] and ensuring power sharing simultaneously [110, 120, 234, 238]. In existing works, microgrid are considered as a multi-agent communication structure, where each DER is an agent. On this basis, the restoration of voltage and frequency is transformed into a consensus tracking problem based on multi-agent collaborative control under leader-following form, and correspondingly, power sharing is seen as an leaderless consensus problem [109, 234, 240]. The centralized control problem of acmicrogrids with a large number of DER agents under constant communication delay has been studied in [242]; in order to further maintain proportional load sharing under time varying delay communication, a distributed control scheme was adopted in [120]; with the help of small-signal analysis methods, [132, 244] respectively designed distributed frequency controllers to obtain the delay margins under the premise of stable AC microgrids; in networks with time varying delays and switching topologies, a leader following distributed control scheme is designed to achieve voltage and frequency recovery while ensuring power sharing [189]; there exists the constant time-delay in communication between DER agents, a robust distributed secondary control strategy is proposed in [106], which increases the network's tolerance for time-delay; immediately after, a distributed control algorithm based on multi-agent was adopted for transmission delay in [249], which eliminated voltage deviation in AC microgrids and maintained current sharing; additionally, on the economic dispatch optimization scheme of AC microgrids, authors in [250] implemented a distributed optimal regulation strategy in the presence of communication delay.

Although there has been some progress in the study of communication delay, communication channels are actually affected by the dual effects of noise and delay in AC microgrids. In terms of the impact of noise on communication links, the sensitivity of microgrid networks to noise can reduce their dynamic performance and even lead to system instability. Therefore, it is urgent to study how to achieve distributed secondary voltage restoration of AC autonomous microgrids under the dual effects of noise and communication delay.

Based on the above discussion, compared to the previous research mentioned, in this chapter, we investigate the stable operation problem of AC microgrids under the influence of measurement noise and time-delay. It will designs a consensus-based distributed control scheme to achieve optimal active power sharing while synchronizing the output voltage and frequency of DER agents. By suppressing the dual interference of noise and time-delay, control designs were implemented for the frequency, voltage, and active power of AC microgrids, ensuring synchronous restoration of voltage and frequency and optimal sharing of active powers. Compared to existing methods [112, 120, 238, 250], we do not need to perform tertiary control layers for economic optimization. The proposed control scheme can achieve economic scheduling of AC microgrids at the secondary control layer, simplifying controller design while reducing time-scale differentiation. It should be noted that unlike existing distributed control methods that adopt mean-square convergence for disturbances [143, 144, 251], a resilient distributed control method will be developed to restore the voltage and frequency of AC microgrids based on obtained the noisy and delayed measurements in almost sure convergence.

6.1 Problem Statement

For an isolated AC microgrid power system composed of N DER agents connected in parallel, the nonlinear dynamics of each DER agent are constructed in its individual direct-quadrature d-q reference frame. Note that each DER agent operates through rotation frequency ω^{com} in the same common reference frame. Therefore, the angle δ_i of DER$_i$ agent can be obtained by

$$\delta_i = \int (\omega_i - \omega^{\mathrm{com}})\mathrm{d}t \tag{6.1}$$

By connecting the d-axis voltage of the DER$_i$ agent to the terminal voltage of the AC microgrid bus, the local droop-based control form can be obtained

$$\begin{cases} \begin{cases} v_i^{\mathrm{od}} = v_i^{\mathrm{nom}} - n_i^Q Q_i \\ v_i^{\mathrm{oq}} = 0 \end{cases} \\ \omega_i = \omega_i^{\mathrm{nom}} - m_i^P P_i \end{cases} \tag{6.2}$$

The entire primary control process adopts a three loop scheme of power, voltage, and current. A virtual leader DER$_0$ was introduced in the tertiary control to specify reference values for voltage v^{DES} and frequency ω^{DES}. In addition, all DER agents interact with their neighbors in a low-bandwidth communication environment through the proposed distributed protocol driver. Finally, the nominal voltage v_i^{nom} and frequency ω_i^{nom} are calculated in real-time through the consensus control strategy, and the output voltage v_i^{o} and frequency ω_i of DER agents are synchronized to the reference value in the power loop based on droop control.

For the power controller, after considering power calculation, low-pass filter, and droop control, the voltage and operating frequency of its inverter bridge can be obtained. According to the cutoff frequency of the low-pass filter ω_i^{c}, the dynamics of the power control loop can be obtained as follows [189].

$$\begin{cases} \dot{P}_i = \dfrac{\omega_i^{\mathrm{c}}}{\omega_i^{\mathrm{c}} + s}[(v_i^{\mathrm{od}} i_i^{\mathrm{od}} + v_i^{\mathrm{oq}} v_i^{\mathrm{oq}}) - P_i] \\ \dot{Q}_i = \dfrac{\omega_i^{\mathrm{c}}}{\omega_i^{\mathrm{c}} + s}[(v_i^{\mathrm{od}} i_i^{\mathrm{oq}} - v_i^{\mathrm{oq}} v_i^{\mathrm{od}}) - Q_i] \end{cases} \tag{6.3}$$

It is worth noting that the nominal frequency ω_i^{nom} of the pulse-width modulation (PWM) inverter bridge is given by the power control loop. At the same time, voltage reference values $v_{i\mathrm{ref}}^{\mathrm{Id}}$ and $v_{i\mathrm{ref}}^{\mathrm{Iq}}$ supplied by the PWM inverter bridge are also provided to the voltage and current control loops. Based on this, the output of the voltage loop is represented as

$$\begin{cases} i_{i\mathrm{ref}}^{\mathrm{Ld}} = k_i^{\mathrm{P}v}(v_{i\mathrm{ref}}^{\mathrm{od}} - v_i^{\mathrm{od}}) + k_i^{\mathrm{I}v} \displaystyle\int (v_{i\mathrm{ref}}^{\mathrm{od}} - v_i^{\mathrm{od}}) \mathrm{d}t - \omega_i^{\mathrm{nom}} C^{\mathrm{f}} v_i^{\mathrm{oq}} + F i_i^{\mathrm{od}} \\ i_{i\mathrm{ref}}^{\mathrm{Lq}} = k_i^{\mathrm{P}v}(v_{i\mathrm{ref}}^{\mathrm{oq}} - v_i^{\mathrm{oq}}) + k_i^{\mathrm{I}v} \displaystyle\int (v_{i\mathrm{ref}}^{\mathrm{oq}} - v_i^{\mathrm{oq}}) \mathrm{d}t + \omega_i^{\mathrm{nom}} C^{\mathrm{f}} v_i^{\mathrm{od}} + F i_i^{\mathrm{oq}} \end{cases} \tag{6.4}$$

where $k_i^{\mathrm{P}v}$, $k_i^{\mathrm{I}v}$ denote the proportional-integral (PI) parameters in voltage loop, respectively; F is the feed-forward gain. With the aid of current reference values, the output of the current loop is represented as

$$\begin{cases} i_{i\mathrm{ref}}^{\mathrm{Id}} = k_i^{\mathrm{P}i}(i_{i\mathrm{ref}}^{\mathrm{od}} - i_i^{\mathrm{od}}) + k_i^{\mathrm{I}i} \displaystyle\int (i_{i\mathrm{ref}}^{\mathrm{od}} - i_i^{\mathrm{od}}) \mathrm{d}t - \omega_i^{\mathrm{nom}} L^{\mathrm{f}} i_i^{\mathrm{Lq}} + v_i^{\mathrm{Id}} \\ i_{i\mathrm{ref}}^{\mathrm{Iq}} = k_i^{\mathrm{P}i}(i_{i\mathrm{ref}}^{\mathrm{oq}} - i_i^{\mathrm{oq}}) + k_i^{\mathrm{I}i} \displaystyle\int (i_{i\mathrm{ref}}^{\mathrm{oq}} - i_i^{\mathrm{oq}}) \mathrm{d}t + \omega_i^{\mathrm{nom}} L^{\mathrm{f}} i_i^{\mathrm{Ld}} + v_i^{\mathrm{Iq}} \end{cases} \tag{6.5}$$

where $k_i^{\mathrm{P}i}$ and $k_i^{\mathrm{I}i}$ denote the PI parameters in current loop.

Next, we can obtain the following dynamic form

$$\begin{cases} \dot{i}_i^{\mathrm{Ld}} = -\dfrac{R_i^{\mathrm{f}}}{L_i^{\mathrm{f}}} i_i^{\mathrm{Ld}} + \omega_i i_i^{\mathrm{Lq}} + \dfrac{1}{L_i^{\mathrm{f}}} v_i^{\mathrm{Id}} - \dfrac{1}{L_i^{\mathrm{f}}} v_i^{\mathrm{od}} \\ \dot{i}_i^{\mathrm{Lq}} = -\dfrac{R_i^{\mathrm{f}}}{L_i^{\mathrm{f}}} i_i^{\mathrm{Lq}} + \omega_i i_i^{\mathrm{Ld}} + \dfrac{1}{L_i^{\mathrm{f}}} v_i^{\mathrm{Iq}} - \dfrac{1}{L_i^{\mathrm{f}}} v_i^{\mathrm{oq}} \\ \dot{v}_i^{\mathrm{od}} = \omega_i v_i^{\mathrm{oq}} + \dfrac{1}{C_i^{\mathrm{f}}} i_i^{\mathrm{Ld}} - \dfrac{1}{C_i^{\mathrm{f}}} i_i^{\mathrm{od}} \\ \dot{v}_i^{\mathrm{oq}} = -\omega_i v_i^{\mathrm{od}} + \dfrac{1}{C_i^{\mathrm{f}}} i_i^{\mathrm{Lq}} - \dfrac{1}{C_i^{\mathrm{f}}} i_i^{\mathrm{oq}} \\ \dot{i}_i^{\mathrm{od}} = -\dfrac{R_i^{\mathrm{o}}}{L_i^{\mathrm{o}}} i_i^{\mathrm{od}} + \omega_i i_i^{\mathrm{oq}} + \dfrac{1}{L_i^{\mathrm{o}}} v_i^{\mathrm{od}} - \dfrac{1}{L_i^{\mathrm{o}}} v_i^{\mathrm{bd}} \\ \dot{i}_i^{\mathrm{oq}} = -\dfrac{R_i^{\mathrm{o}}}{L_i^{\mathrm{o}}} i_i^{\mathrm{oq}} + \omega_i i_i^{\mathrm{od}} + \dfrac{1}{L_i^{\mathrm{o}}} v_i^{\mathrm{oq}} - \dfrac{1}{L_i^{\mathrm{o}}} v_i^{\mathrm{bq}} \end{cases} \tag{6.6}$$

Integrating (6.2)–(6.6), the signaling model of the entire AC microgrid is represented as

$$\begin{cases} \dot{x}_i = f_i(x_i) + g_i(x_i)u_i \\ y_i = h_i(x_i) \end{cases} \tag{6.7}$$

where $x_i = [P_i, Q_i, i_i^{\mathrm{Ld}}, i_i^{\mathrm{Lq}}, v_i^{\mathrm{od}}, v_i^{\mathrm{oq}}, i_i^{\mathrm{od}}, i_i^{\mathrm{oq}}]$ denotes the AC microgrid state vector, and the specific expressions of $f_i(x_i)$, $g_i(x_i)$ and $h_i(x_i)$ can be analyzed from (6.1)–(6.6), so this process is omitted here. It can be clearly seen that $g_i(x_i)$ does not include any system parameters, so it can be regarded as a constant matrix; and $f_i(x_i)$ represents the whole network nonlinear dynamics; and $h_i(x_i)$ is a constant matrix, obviously, for convenience in subsequent expressions, it will be abbreviated as h.

Select input variables in (6.2) to design controllers for v_i^{od} and ω_i respectively, in order to establish resilient distributed consensus control schemes that synchronizes all DERs' outputs to reference values v_{ref} and ω_{ref} of AC microgrids under the dual effects of noise and time delays.

6.2 Resilient Distributed Cooperative Intelligence

In this section, resilient distributed control strategies were designed for the voltage/frequency and optimal power-sharing of the AC microgrid network, including leader-following and leaderless approaches. By designing three distributed controllers to achieve synchronous reference output of voltage and frequency and active power sharing of DER agents. Concretely, according to the distributed control protocol, adjust v_i^{nom} and ω_i^{nom} in (6.2) to achieve synchronous output of terminal voltage v_i^{od}/frequency ω_i^{nom} to reference values v^{DES} and ω^{DES}. Given that the output voltage is $v_i = \sqrt{(v_i^{\mathrm{od}})^2 + (v_i^{\mathrm{oq}})^2}$, it is necessary to select the nominal point v_i^{nom} such that $v_i^{\mathrm{od}} \to v^{\mathrm{DES}}$ can achieve voltage restoration $v_i \to v^{\mathrm{DES}}$ in (6.2). Similar design ideas can make $\omega_i \to \omega^{\mathrm{DES}}$ hold by selecting ω_i^{nom}. The power sharing among DER agents is achieved through a leaderless protocol driver.

The successful implementation of the above control design requires each DER agent to have communication, data processing, and decision-making capabilities. To this end, a low-bandwidth communication network is introduced to describe the entire dynamic process, which includes the characteristics of the AC microgrid without loss of generality. It is worth noting that compared to general communication network structures, it has a simpler communication structure, more flexible expansion methods, and more reliable transmission processes [106, 240, 249].

6.2.1 Frequency Restoration

A distributed protocol with leader-following is established to solve the problem of synchronizing frequencies of all DER agents to the frequency reference ω^{DES}. It should be pointed out in advance that we assume each DER agent can exchange information through a low-bandwidth network that is subject to disturbances. According to the form of (6.2), the frequency dynamics of DER$_i$ agent can be further obtained as follows:

$$\dot{\omega}_i = \dot{\omega}^{\mathrm{nom}} - m_i^P \dot{P}_i = u_i^{\omega} \tag{6.8}$$

with the controller for frequency u_i^{ω} is designed in the following form

$$\begin{aligned} u_i^{\omega} =& c(t) \sum_{j \in \mathcal{N}_i} a_{ij} \left[\omega_j(t-\tau) - \omega_i(t-\tau) + \sigma_{ij} \varpi_{ij} \right] \\ &+ c(t) a_{i0} \left[\omega^{\mathrm{DES}} - \omega_i(t-\tau) + \sigma_i \varpi_i \right] \end{aligned} \tag{6.9}$$

which makes $\lim_{t\to\infty} E \left\| \omega_i(t) - \omega^{\mathrm{DES}} \right\| = 0$. The result here is referred to as almost sure convergence through the dual suppression of noise and time-delay by the distributed frequency controller (6.9), where $c(t)$ is the time varying parameter satisfying [252]

$$\begin{cases} \int_0^{\infty} c(s)\mathrm{d}s = \infty \\ \int_0^{\infty} c^2(s)\mathrm{d}s < \infty \\ \lim_{t\to\infty} c(t) = 0 \end{cases} \tag{6.10}$$

And $c(t)$ has a upper bound $\bar{c} = \sup_{t\geq 0} c(t)$, where $\bar{c}$ is a constant greater than zero. In fact, during the communication process, the network is very sensitive to additive noise, which will have a negative impact on the achievement of control objectives [164]. A specific example is the additive noise generated by the receiver front-end and the environmental noise received by the antenna during wireless communication. In addition, different environmental conditions can also lead to unavoidable noise, such as rainwater causing noise in the channel. Note that the additive noise at the receiving end, electronic components, amplifiers, and other devices is unified as thermal noise, which is essentially Gaussian noise [120]. Thereby, the designed controller takes into account the issue of noise suppression, $\tau > 0$ denotes the symmetric constant communication delay; ϖ_{ij} represent independent Gaussian white noise; and $\varpi_{ij} \geq 0$ expresses the noise intensity. Before analyzing the impact of the designed distributed frequency controller on the AC microgrid system, the following symbols are defined.

Denote the vectors $\omega = [\omega_1, \omega_2, \cdots, \omega_N]^T \in R^N$, $\omega^{\mathrm{DES}} = 1_N \otimes \omega^{\mathrm{DES}} \in R^N$, $\Lambda_0 = \mathrm{diag}\{\sigma_1, \sigma_2, \cdots, \sigma_N\}$, $\Lambda_i = \mathrm{diag}\{\sigma_{i1}, \sigma_{i2}, \cdots, \sigma_{iN}\}$ with $i = 1, \cdots, N$, $\varpi_0 = [\varpi_1, \varpi_2, \cdots, \varpi_N]^T \in R^N$, $\varpi_i = [\varpi_{i1}, \varpi_{i2}, \cdots, \varpi_{iN}]^T \in R^N$, $\varpi_0 = [\varpi_0^T, \varpi_1^T, \cdots, \varpi_N^T]^T \in R^{N(N+1)}$, $a(i, \cdot)$ express the ith row of $A = (a_{ij})_{N\times N}$, which makes a matrix

$$\Lambda = \begin{pmatrix} a_{10}\Lambda_1 & & & a(1,\cdot)\Lambda_1 & & \\ & \ddots & & & \ddots & \\ & & a_{N0}\Lambda_N & & & a(N,\cdot)\Lambda_N \end{pmatrix}$$

where $\Lambda \in R^{N\times(N(N+1))}$. Rewrite (6.8) and (6.9) to obtain the following compact form

$$\omega = -c(t)(L + B)\omega(t - \tau) + c(t)B\omega^{\mathrm{DES}} + \Lambda\varpi \tag{6.11}$$

Furthermore, the stochastic form of (6.11) can be obtained

$$\mathrm{d}\omega = c(t)[-(L + B)\omega(t - \tau) + B\omega^{\mathrm{DES}}]\mathrm{d}t + c(t)\Lambda\mathrm{d}\varpi \tag{6.12}$$

According to the designed distributed frequency control protocol (6.9), the following conclusions are valid.

Theorem 6.1 *For the directed spanning tree graph $\hat{G}$ with root node r, and $a_{r0} > 0$ $\forall r \in \{1, \cdots, N\}$. For $\tau < \tau^* = (1/\lambda_N\bar{c})$, distributed controllers* (6.9), (6.33) *and* (6.37) *can synchronize the output voltage v_i and frequency ω_i of DER agents to reference values v^{DES} and ω^{DES} in almost sure convergence, while ensuring optimal active power sharing.*

Proof Define the following error coordinate transformation

$$e(t) = \omega(t) - \omega^{\mathrm{DES}} \tag{6.13}$$

Furthermore, the stochastic differential delay equation of the closed-loop AC microgrid with additive noise can be obtained as follows:

$$\mathrm{d}e = -c(t)(L + B)e(t - \tau)\mathrm{d}t + c(t)\Lambda\mathrm{d}\varpi \tag{6.14}$$

The fact is that $L + B$ is symmetrical since $\hat{G}$ is undirected, all eigenvalues of ($\boldsymbol{L}$ + $\boldsymbol{B}$) are positive [163]. Thus, there is a matrix Q which makes $Q(L + B)Q^{-1} = \mathrm{diag}(\lambda_1, \lambda_2, \cdots, \lambda_N) = \Xi$. Define $\delta(t) = Qe(t)$, with the aid of Ito formula, one has

$$\begin{aligned} \mathrm{d}\delta &= Q\mathrm{d}e(t) \\ &= -c(t)Q(L + B)e(t - \tau)\mathrm{d}t + c(t)Q\Lambda\mathrm{d}\varpi \\ &= -c(t)Q(L + B)Q^{-1}\delta(t - \tau)\mathrm{d}t + c(t)Q\Lambda\mathrm{d}\varpi \\ &= -c(t)\Xi\delta(t - \tau)\mathrm{d}t + c(t)Q\Lambda\mathrm{d}\varpi \end{aligned} \tag{6.15}$$

Furthermore, it can be obtained from (6.15) that

$$\mathrm{d}\delta_k = -\lambda_k c(t)\delta_j(t-\tau) + c(t)Q_k\Lambda\mathrm{d}\varpi \tag{6.16}$$

where $k = 1, \cdots, N$; $\delta(t) = [\delta_1^T(t), \cdots, \delta_{N1}^T(t)]^T$ with $\delta_N^T(t) \in R^n$; and Q_k denotes the kth row of the matrix Q.

Note that it is necessary to ensure that the established control protocol is almost sure convergence under any initial conditions $\omega(0)$, i.e., $\lim_{t\to\infty} \|\delta_k(t)\| = 0, \forall k = 0, \cdots, N$. Define the solution of (6.16) is $\pi_{k1}(t)$ for $\tau = 0$, while $\pi_{k2}(t)$ for $\tau \neq 0$, respectively.

After Eqs. (2.9) and (6.10) hold true, with the help of Lemma 2.5, it can be obtained that

$$\lim_{t\to\infty} |\pi_{k1}(t)\| = 0 \tag{6.17}$$

Define $\pi_k(t) = \pi_{k1}(t) - \pi_{k2}(t)$, which implies $\pi_j(0) = 0$ and for $t > 0$, it has

$$\dot{\pi}_k(t) = -\lambda_k c(t)\pi_k(t-\tau) + c(t)\lambda_k[\pi_{k1}(t-\tau) - \pi_{k1}(t)] \tag{6.18}$$

Since $\lim_{k\to\infty} \|\pi_{ki}(t)\| = 0$, one has

$$\lim_{k\to\infty} \|\pi_{ki}(t-\tau) - \pi_{ki}(t)\| = 0 \tag{6.19}$$

By using the variation of the constants formula, (6.18) can be transformed to

$$\pi_k(t) = \int_0^t \vartheta_k(t,s)c(s)\lambda_k[\pi_{k1}(s-\tau) - \pi_{k1}(s)]\mathrm{d}s \tag{6.20}$$

where $\vartheta_k(t,s)$ represents the differential resolvent function for $\dot{Y}(t) = -\lambda_k c(t)Y(t-\tau)$.

Go a step further, one can obtain that

$$\begin{aligned} \|\pi_k(t)\| &\leq \int_0^t \|\vartheta_k(t,s)\| \, \|c(s)\| \, \|\lambda_k[\pi_{k1}(s-\tau) - \pi_{k1}(s)]\| \, \mathrm{d}s \\ &\leq \mu_0^{0.5} \int_0^t e^{0.5\eta_0 \int_s^t c(u)\mathrm{d}u} c(s) \, \|\lambda_k[\pi_{k1}(s-\tau) - \pi_{k1}(s)]\| \, \mathrm{d}s \end{aligned} \tag{6.21}$$

Then we denote

$$\phi(t) = \int_0^t e^{0.5\eta_0 \int_s^t c(u)\mathrm{d}u} c(s) \, \|\lambda_k[\pi_{k1}(s-\tau) - \pi_{k1}(s)]\| \, \mathrm{d}s \tag{6.22}$$

and

$$\Phi(t) = e^{0.5\eta_0 \int_s^t c(u)\mathrm{d}u}\phi(t) \tag{6.23}$$

which implies $\phi(t)$ is increasing, it means $\lim_{t\to\infty}\phi(t) = \infty$ or $\lim_{t\to\infty}\phi(t) < \infty$.

The obvious conclusion can be drawn that $\lim_{t\to\infty}\|\Phi(t)\| = 0$ for the condition $\lim_{t\to\infty}\phi(t) < \infty$ in (2.9). Otherwise, $\lim_{t\to\infty}\phi(t) = \infty$, and with the aid of L'Hospital's rule yields

$$\begin{aligned}\lim_{x\to\infty}\|\Phi(t)\| &= \frac{\int_0^t e^{0.5\eta_0 \int_s^t c(u)\mathrm{d}u} c(s)\,\|\lambda_k[\pi_{k1}(s-\tau) - \pi_{k1}(s)]\|\,\mathrm{d}s}{e^{0.5\eta_0 \int_s^t c(u)\mathrm{d}u}} \\ &= \mu_0^{0.5}\lim_{x\to\infty}\|\lambda_k[\pi_{k1}(s-\tau) - \pi_{k1}(s)]\| = 0\end{aligned} \tag{6.24}$$

The conclusion that can be summarized is $\lim_{t\to\infty}\|\pi_k(t)\| = 0$. Go a step further, according to $\lim_{t\to\infty}\|\pi_{k1}(t)\| = 0$, one has $\lim_{t\to\infty}\|\pi_{k2}(t)\| = 0$, the output frequency ω_i of DER agents can synchronize to the reference value ω^{DES} in almost sure convergence. Partial proof has been provided.

6.2.2 *Optimal Power Sharing*

A leaderless distributed consensus controller will be designed in this section to achieve optimal power sharing among DER agents. Introduce the cost functions of DER agents as follows [253]

$$f_i(P_i) = \alpha_i P_i^2 + \beta_i P_i + \gamma_i \tag{6.25}$$

where α_i, β_i and γ_i denote the cost weights of DER$_i$ agent; and P_i represents the output active power of DER$_i$ agent.

Considering the actual supply-demand balance of AC microgrids, the total cost $f_i(P_i)$ of electricity generation can take the following form

$$\begin{aligned}&\min f(P) = \sum_{i=1}^{n} f_i(P_i) \\ &\text{s.t.} \sum_{i=1}^{n} f_i(P_i) = P_D\end{aligned} \tag{6.26}$$

where P_D denotes the minimum power requirement that all DER agents in AC microgrids need to satisfy.

Establish the following Lagrange function to solve the optimal problems established in (6.25) and (6.27).

$$L_\alpha = \sum_{i=1}^{n} f_i(P_i) + \eta \left(P_0 - \sum_{i}^{n} P_i \right) \tag{6.27}$$

where η represents the Lagrange multiplier. By solving along the direction of optimal power, we can obtain

$$\frac{\partial L_\alpha}{\partial i_i} = \frac{\partial C_i(i_i)}{\partial i_i} - \eta_i = 2\alpha_i i_i + \beta_i - \eta_i \tag{6.28}$$

The form of the incremental cost function is

$$\eta_i(i_i) = 2\alpha_i i_i + \beta_i \tag{6.29}$$

A necessary and sufficient condition for minimizing the generation cost of DER agents to optimize economic operation while ensuring supply-demand balance is that the incremental cost of all DER agents needs to be unified [163], that is

$$\eta_1(i_i) = \cdots = \eta_i(i_i) = \cdots = \eta_N(i_i) \tag{6.30}$$

When AC microgrids approach the steady state, frequencies ω_i of all DER agents tend to be equal, so all $\eta_i(P_i)$ are also equal, enabling optimal active power sharing of DER agents [253]. Next, it is necessary to ensure the active power sharing of each DER agent, and the necessary and sufficient condition that the incremental cost of each DER agent is equal can be rewritten as

$$\eta_1(P_1) = \cdots = \eta_i(P_i) = \cdots = \eta_N(P_N) \tag{6.31}$$

In order to manage all DER agents with different differences and achieve active power sharing, a responsive leaderless distributed power control protocol needs to be designed. The dynamics of active power can be obtained through (6.29) as follows:

$$\dot{\eta}_i(P_i) = u_i^P \tag{6.32}$$

with

$$u_i^P = c(t) \sum_{j \in \mathcal{N}_i} a_{ij} \left[\eta_j \left(P_j(t-\tau) \right) - \eta_i \left(P_i(t-\tau) \right) + \sigma_{ij} \varpi_{ij} \right] \tag{6.33}$$

which demonstrates that

$$\lim_{t \to +\infty} E \left\| \eta_i P_i(t) - \eta_j P_j(t) \right\| = 0, \ \forall i \neq j \in \{1, 2 \cdots, N\} \tag{6.34}$$

Driven by the leader-following distributed frequency control protocol (6.9) and leaderless distributed active power control protocol (6.33), the droop-based primary control process is represented as

$$\omega_i^{\mathrm{nom}} = \int \left(u_i^{\omega} + m_i^P u_i^P \right) \mathrm{d}t \tag{6.35}$$

It can be clearly seen that the leaderless distributed active power controller (6.33) lacks the second term of the leader-following distributed frequency controller (6.9) in form. Therefore, for Theorem 6.1, it can be modified controller (6.33) can satisfy controller (6.9) when there is no specific synchronization target.

6.2.3 Voltage Regulation

In this section, we further design the leader-following voltage controller to manage the voltage output synchronization of AC microgrids. The dynamic equation for voltage management obtained through (6.2) is

$$\dot{v}_i = \dot{v}_i^{\mathrm{nom}} - n_i^Q \dot{Q}_i = u_i^v \tag{6.36}$$

with the voltage controller are designed as follows

$$\begin{aligned} u_i^v =& c(t) \sum_{j \in \mathcal{N}_i} a_{ij} \left[v_j(t-\tau) - v_i(t-\tau) + \sigma_{ij} \varpi_{ij} \right] \\ & + b_i \left[v^{\mathrm{DES}} - v_i(t-\tau) + \sigma_i \varpi_i \right] \end{aligned} \tag{6.37}$$

which illustrates that

$$\lim_{t \to +\infty} E \left\| v_i(t) - v^{\mathrm{DES}} \right\| = 0, \ \forall i \in \{1, 2 \cdots, N\} \tag{6.38}$$

Thus, the droop-based primary control process is

$$v_i^{\mathrm{nom}} = \int \left(u_i^v + n_i^Q \dot{Q}_i \right) \mathrm{d}t \tag{6.39}$$

The relevant proof process is similar to Sect. 6.2.1, so it is omitted here.

6.2.4 Control Implementation

The proposed resilient distributed multi-agent control scheme consists of three consensus-based control protocols for frequency synchronization, voltage regulation, and optimal active power sharing. These protocols are designed to operate under communication noise and time-delay disturbances, achieving almost sure convergence. The detailed implementation of the proposed distributed intelligent cooperative protocol can be represented as follows:

Step1 Initialization: Set the reference trajectory v^{DES} and ω^{DES}, the initial inputs, $u_i^v(0)$, $u_i^\omega(0)$, and $u_i^P(0)$, and the initial states, $v_i^{\mathrm{od}}(0)$, $\omega_i(0)$, and $P_i(0)$ for all $i = 1, 2, \cdots, N$. Define the adjacency matrix A and the leader adjacency matrix B. Choose the consensus gain function $c(t)$.

Step2 Information processing: Accept the neighboring DER$_j$ states with communication noise and time delay disturbances $v_j(t-\tau)+\sigma_{ij}\varpi_{ij}$, $\omega_j(t-\tau)+\sigma_{ij}\varpi_{ij}$ and $\eta_j P_j(t-\tau)+\sigma_{ij}\varpi_{ij}$. If $b_i > 0$, the DER$_i$ can also receive the reference information from the virtual DER $v^{\mathrm{DES}} + \sigma_i\varpi_i$ and $\omega^{\mathrm{DES}} + \sigma_i\varpi_i$.

Step3 Calculate the secondary control inputs according to (6.9), (6.33) and (6.37).

Step4 Calculate nominal set points: Apply $u_i^v(t)$, $u_{i,k}^\omega(t)$, and $u_{i,k}^P(t)$ to compute $v_i^{\mathrm{nom}}(t)$ and $\omega_i^{\mathrm{nom}}(t)$, for $i = 1, 2, \cdots, N$.

Remark 6.1 In many existing works, only the impact of communication delay [106, 120, 132, 189, 242, 244, 249, 250] or measurement noise [144] on AC microgrids has been considered. However, this chapter considers both of these issues simultaneously due to the high sensitivity of actual communication networks to noise and communication delay, which is a practical and meaningful problem. Simultaneously considering these two issues involves coupling modeling of noise and communication delay, which is a more complex and challenging task. Therefore, the proposed novel control scheme has a good suppression effect on noise and communication delay, and can achieve time frequency, voltage recovery, and optimal active power sharing.

Remark 6.2 In the existing disturbances suppression design work for AC microgrids, the results are all based on the mean square consensus convergence [143, 144, 163, 251]. Nevertheless, in reality, in many applications, there can only be one operating environment in random experiments, so the results in a probabilistic sense are more universal. Based on this, the results obtained from the elastic distributed control strategy proposed in this chapter are almost sure convergence. Moreover, unlike small signal stability analysis [132, 242, 244], the large signal analysis method adopted in this chapter enables us to obtain results with a wider range of applications.

Remark 6.3 In low voltage AC microgrids, the existence of impedance in distribution lines makes it impossible to simultaneously meet voltage regulation and reactive power sharing. The nominal regulation of voltage makes it impossible to balance precise reactive power sharing. Therefore, in this chapter, only the frequency and voltage recovery problems and the optimal sharing of active power in AC microgrids were considered under additive noise and time delay disturbances while the issue of reactive power will be studied in future work. At the same time, it should be pointed out that the designed distributed algorithm is likely to be extended to reactive power sharing under noise and time delay disturbances, therefore, this has the potential for challenges. However, due to limitations in job responsibilities, this will be further developed in future work.

Remark 6.4 The designed time varying consensus gain $c(t)$ plays a pivotal role in attenuating the adverse effects of measurement noise while ensuring convergence. Its allows the system to dynamically adjust the control effort in response to real-time noise levels and communication delays. This adaptability is particularly advantageous in microgrid deployments, where communication quality may fluctuate due to environmental factors or hardware limitations. Unlike existed fixed gain strategies, the proposed gain function enhances robustness without requiring frequent retuning, thereby supporting autonomous operation of microgrids.

Remark 6.5 The stability criteria derived in Theorem 5.1 explicitly depend on the algebraic connectivity of the communication graph. This highlights the importance of designing communication networks with sufficient redundancy and connectivity, especially in islanded microgrids where certain DERs may act as critical information hubs. In practice, the proposed method can be integrated with network reconfiguration strategies to maintain a spanning tree even under link failures or DoS attacks, thereby preserving system stability under dynamic or adversarial conditions.

Remark 6.6 Traditionally, economic dispatch is handled at the tertiary control level with slow time scales. By embedding optimal active power sharing directly into the secondary control layer, the proposed method reduces both computational overhead and communication latency. This integration not only simplifies the hierarchical control structure but also enables faster response to load variations and generation changesa crucial feature for islanded microgrids with high penetration of intermittent renewable sources.

Remark 6.7 While mean square convergence is commonly adopted in stochastic control literature, the convergence established in this work provides a stronger and more practical guarantee. In real microgrids' operation, the secondary controller cannot observer an ensemble average. Thus, the almost sure convergence ensures that the frequency and voltage restoration objectives can be finally realized in presence of the noise and delay processes, enhancing the reliability and applicability of the proposed control strategy in implementations.

6.3 Simulation Results and Discussion

To thoroughly validate the effectiveness of the proposed resilient distributed intelligent cooperative protocol, a series of distributed control tests were carried out within an islanded AC microgrid environment. The test microgrid is configured with five DER agents, as referenced in [164]. These tests were implemented using the MATLAB/SimPowerSystems platform. All component parameters associated

Table 6.1 The parameters values for Microgrid test system

DERs	Parameter	$DER_{1\&4}$ (65 kVA)	Parameter	$DER_{2\&3\&5}$ (63 kVA)
	V^{DC}	800 V	800 V	780 V
	m^P	1.2×10^{-4}	m^P	0.8×10^{-4}
	n^Q	2.4×10^{-5}	n^Q	1.6×10^{-5}
	R^f	0.1 Ω	R^f	0.1 Ω
	C^f	25 μF	C^f	25 μF
	R^o	0.001 Ω	R^o	0.001 Ω
	L^o	0.2 mH	L^o	0.2 mH
	K^{PV}	0.35	K^{PV}	0.01
	K^{IV}	200	K^{IV}	400
	K^{PC}	0.35	K^{PC}	0.01
	K^{IC}	200	K^{IC}	400
Loads	$Load_1$	$Load_2$	$Load_3$	$Load_4$
	P_1 12 kW	P_2 18 kW	P_3 12 kW	P_4 36 kW
Lines	$Line_1$	$Line_2$	$Line_3$	$Line_3$
	R_1 4 Ω	R_2 2.24 Ω	R_3 1.2 Ω	R_4 1.2 Ω
	L_1 1.95 H	L_2 0.79 H	L_3 0.3 H	L_4 0.3 H

with the DER agents are adopted directly from the specifications detailed in [164], which is depicted in Table 6.1.

A fundamental assumption underpinning these experiments is that each DER agent can communicate with its neighboring agents via low bandwidth communication networks. This capability is critical for enabling cooperative control and maintaining system stability, particularly under varying operational conditions. Both the physical connections and the cyber layer topology are illustrated in Fig. 6.1. It is important to emphasize that the physical and cyber topologies are designed with notable structural differences. Such a design reflects practical microgrid implementations, where communication pathways do not necessarily mirror the physical wiring layout. The relevant parameters for graph connections are as follows

$$A = \begin{bmatrix} 0\,1\,0\,0\,0 \\ 1\,0\,1\,0\,0 \\ 0\,1\,0\,1\,0 \\ 0\,0\,1\,0\,1 \\ 0\,0\,0\,1\,0 \end{bmatrix}$$

$$B = \begin{bmatrix} 1\,0\,0\,0\,0 \\ 0\,0\,0\,0\,0 \\ 0\,0\,0\,0\,0 \\ 0\,0\,0\,0\,0 \\ 0\,0\,0\,0\,0 \end{bmatrix}$$

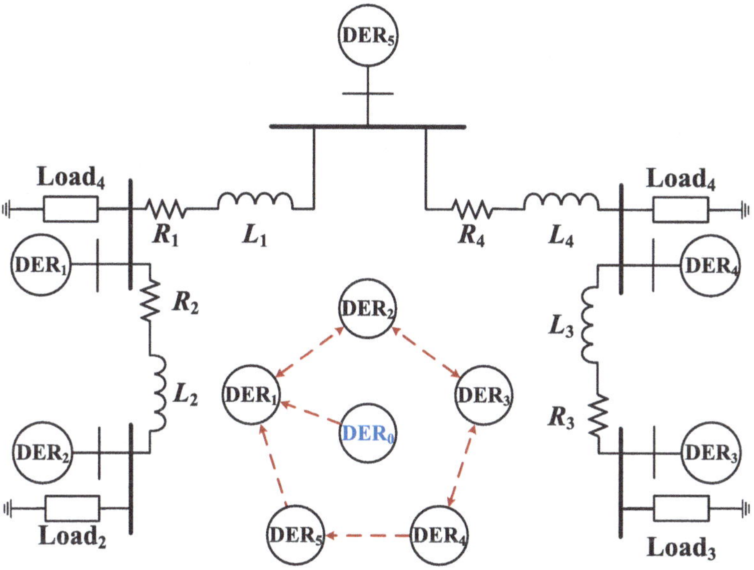

Fig. 6.1 Test AC microgrids: (**a**) physical microgrid test system; (**b**) communication topology G

The stochastic disturbance intensity is set to $\sigma_{ji} = 0.5$ with $i, j \in \{1, 2, 3, 4, 5\}$. The time varying consensus parameter is $c(t) = (1/10(t+1))$, thereby $\bar{c} = \sup_{t \geq 0} c(t) = 0.1$, note that the selected $c(t)$ satisfies condition (2.9). The delay parameter is selected as $\tau = 0.1$. Through calculation, we obtain the eigenvalues of the associated matrix as $\lambda_1 = 0.081$, $\lambda_2 = 0.6903$, $\lambda_3 = 1.7154$, $\lambda_4 = 2.8308$, $\lambda_5 = 3.6825$. Among these, the largest eigenvalue λ_5 plays a critical role in determining stability. Specifically, the product $\tau \lambda_5 \bar{c}$ evaluates to 0.0368, which is strictly less than 1. This condition confirms that the chosen delay $\tau = 0.1$ adheres to the requirements set forth in Theorem 6.1, thereby ensuring the desired convergence properties of the system.

6.3.1 *Robustness Against Complex Operation Ability*

In this test example, the proposed distributed frequency (6.9), voltage (6.37), and active power (6.33) controllers are in operation throughout the entire process. Test the noise resiliency of the proposed control scheme on the entire AC microgrid through the following dynamic process.

(1) At $t = 0$ s. The AC microgrid operates in islanded modes.
(2) At $t = 7.5$ s. Load_1 and Load_3 are set to zero.
(3) At $t = 15$ s. Load_1 and Load_3 restored to full load.
(4) At $t = 22.5$ s. DER_5 is plugged out from the AC microgrid.

(5) At $t = 30$ s. DER_5 is plugged back into the AC microgrid.

The experimental results associated with frequency restoration, voltage regulation, and optimal active power sharing under various time-delay conditionsnamely, no delay, small delay, and large delayare illustrated in Figs. 6.2, 6.3, and 6.4, respectively. As observed in Figs. 6.2a, b and 6.3a, b, when the AC microgrid operates in islanded mode, both the output voltage and frequency of all distributed energy resource (DER) agents synchronize accurately to their reference values, specifically $v^{DES} = 380$ V and $\omega^{DES} = 314$ rad/s. This synchronization is maintained even during subsequent variations in load and throughout plug-and-play operations. In such dynamic scenarios, all DERs exhibit only brief transient deviations before re-synchronizing to their prescribed reference values.

Furthermore, the active power outputs of the DER agents consistently form two distinct clusters. This clustering stems from the underlying cost functions, which can drive higher cost DERs to adjust their power outputs so that lower cost DERs will assume a greater proportion of the entire load in the test microgrid system. As a result, the system can finally achieve an economically efficient distribution of active power, aligning with the objective of optimal power sharing.

As depicted in Figs. 6.2c and 6.3c, at $t = 22.5$ s, the DER_5 agent is plugged out from AC microgrids. The power previously supplied by this DER is then redistributed among the remaining DER agents. Driven by the active power controller, the output of the other four DERs increases accordingly. Notably, this redistribution does not compromise optimal active power distribution. The system continues to maintain optimal active power sharing among the operational DERs. Conversely, when the DER_5 is reconnected to the microgrid at $t = 30$ s, all DER agents rapidly revert to their original power sharing states. This demonstrates the robustness and adaptability of the control strategy in presence of DERs' plug-and-play. Overall, the results confirm that the proposed control approach ensures reliable frequency and voltage regulation, as well as economically optimal power sharing, under various operational conditions and even in the presence of communication time delays.

A comparative analysis of the results presented in Figs. 6.2 and 6.3 reveals significant differences in both convergence speed and transient overshoot across the tested scenarios. These performance variations can be attributed to the presence of noise and communication time delays in the communication network, which are known to significantly degrade the control performance of AC microgrids. Specifically, noise and communication time delays will disrupt the coordination among DER agents, thereby prolonging the settling time and amplifying overshoot during transients. It is important to emphasize that the experimental results clearly demonstrate that the synchronization in frequency restoration and voltage regulation, as well as optimal active power sharing can only be achieved when the communication time delays and noise levels remain within the bounds specified in Theorem 6.1. These limiting conditions act as a critical threshold for maintaining microgrid stability and control accuracy. Once these constraints are violated, the cooperative control performance deteriorates considerably.

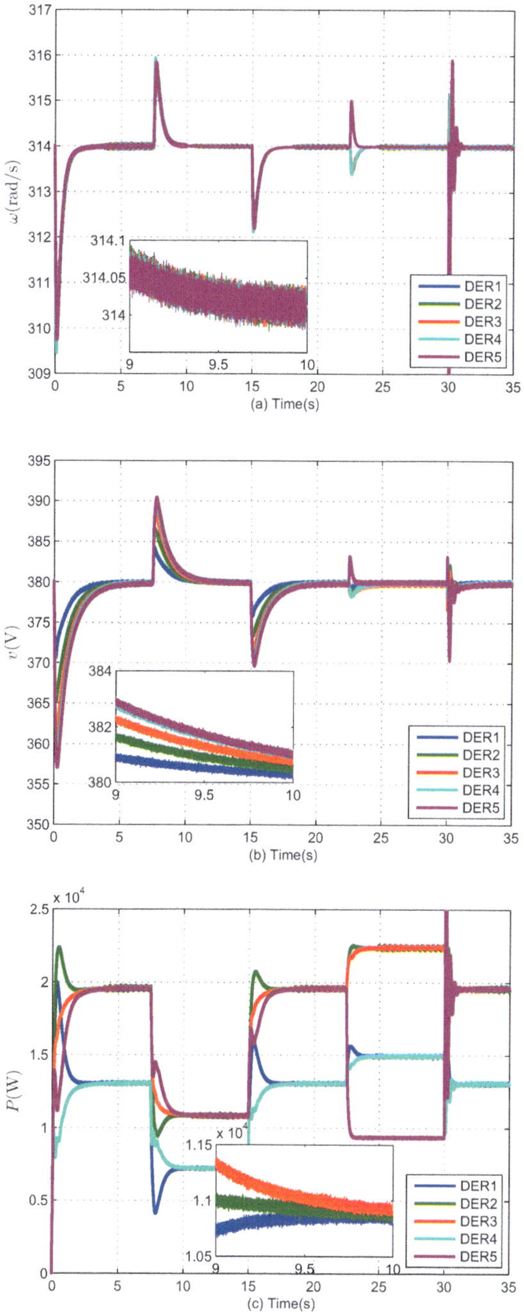

Fig. 6.2 Almost sure synchronization performance: (**a**) frequency; (**b**) output voltage; and (**c**) optimal active power, subject to additive noises (©2024 IEEE. Reprinted, with permission, from IEEE Transactions on Systems, Man, and Cybernetics: Systems [164])

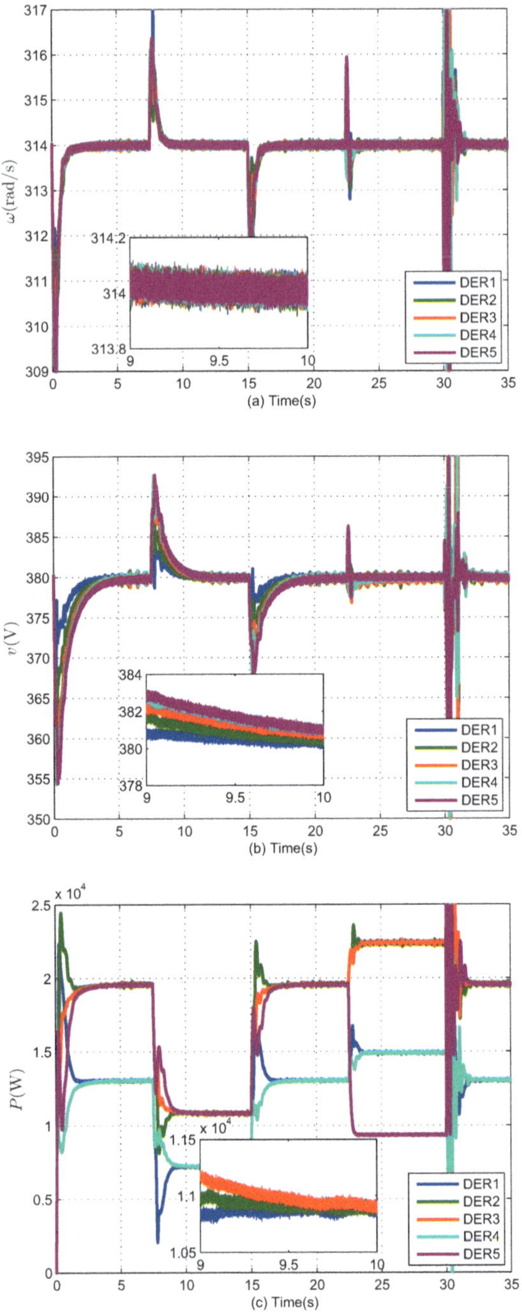

Fig. 6.3 Almost sure synchronization performance: (**a**) frequency; (**b**) output voltage; and (**c**) optimal active power, subject to additive noise and small time delay disturbances (©2024 IEEE. Reprinted, with permission, from IEEE Transactions on Systems, Man, and Cybernetics: Systems [164])

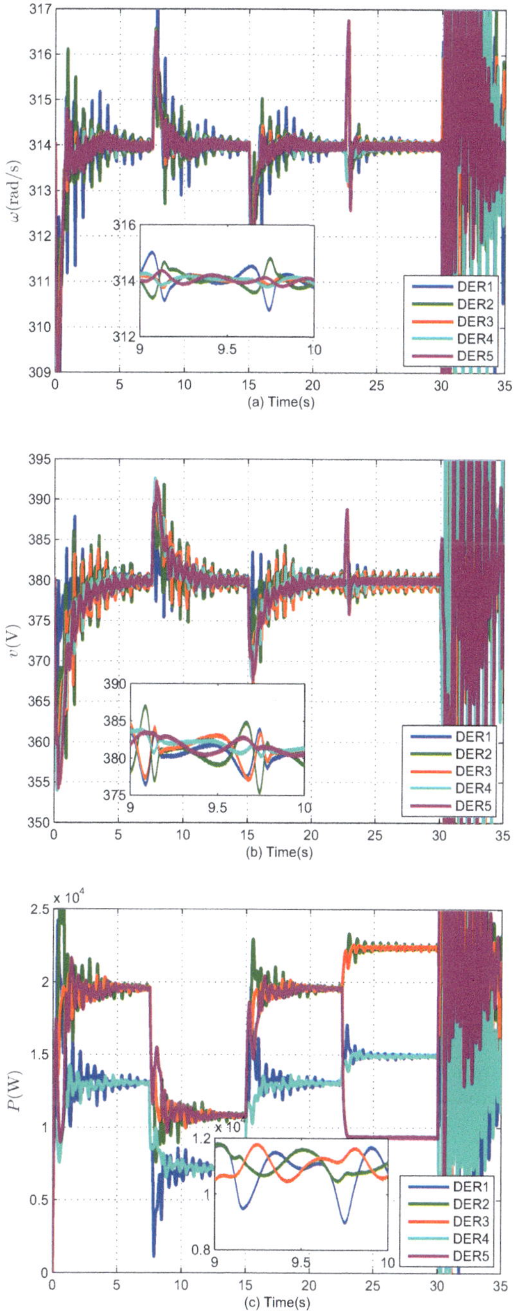

Fig. 6.4 Almost sure synchronization performance: (**a**) frequency; (**b**) output voltage; and (**c**) optimal active power, subject to additive noise and large time delay disturbances (©2024 IEEE. Reprinted, with permission, from IEEE Transactions on Systems, Man, and Cybernetics: Systems [164])

This point can further be illustrated by the simulation results depicted in Fig. 6.4, where the communication time delays are set to $\tau = 10$. Under this configuration, the value $\tau\lambda_6\bar{c} = 3.68 > 1$ exceeds the unity bound required by Theorem 6.1. As a consequence, the microgrid system fails to meet the secondary control objectives: both frequency and voltage deviate persistently from their reference values, and the active power sharing among DERs becomes suboptimal. The loss of synchronization and poor active power distribution in this case illustrate the importance of constraints of communication time delays derived in the corresponding theoretical analysis. In summary, these simulation results validate that the stability and performance of the proposed distributed resilient seconday control strategy are highly sensitive to communication time delays, and the conditions established in Theorem 6.1 serve as a necessary guarantee for reliable microgrid operation.

6.3.2 Comparison with Previously-Proposed Methods

In this comparative study, the performance of the proposed resilient distributed secondary control scheme is investigated against existing control methods to further demonstrate its superior capabilities. Following the same simulation procedure adopted in the previous analysis, the secondary control strategy introduced in [189] under identical operating conditions is implemented. The corresponding simulation results are represented in Fig. 6.5, which illustrate significant limitations of the earlier method when subjected to communication time delays and noises.

As shown in Fig. 6.5, when an additive noise with intensity $\theta_{ij} = 0.1$ is introduced at $t = 1$ s, the test microgrid system under the secondary control method from [189] exhibits sustained oscillations in curves of both operating frequencies and output voltage amplitudes. These oscillations also exist in the optimal active power output curves, which prevent the system from reaching a steady state operation. Over time, the growing amplitude of these deviations leads to a complete loss of stability, which demonstrate the existing method's limitation of robustness in noisy environments.

In contrast, the proposed resilient distributed secondary control approach proposed in this chapter maintains excellent control performance under the same challenging conditions. As depicted in Fig. 6.2, the proposed control scheme can effectively suppress the adverse effects of additive noise and communication time delays. The system achieves smooth convergence without significant overshoot or oscillatory behavior, ensuring reliable restoration of frequency and voltage to their reference values while maintaining optimal active power sharing.

The marked difference in performance can be attributed to the inherent robustness of the proposed control design in this chapter, which incorporates specific mechanisms to attenuate noise and compensate for communication time delays. Whereas conventional secondary control methods like that in [189] prove vulnerable to such disturbances, the proposed resilient secondary control approach demon-

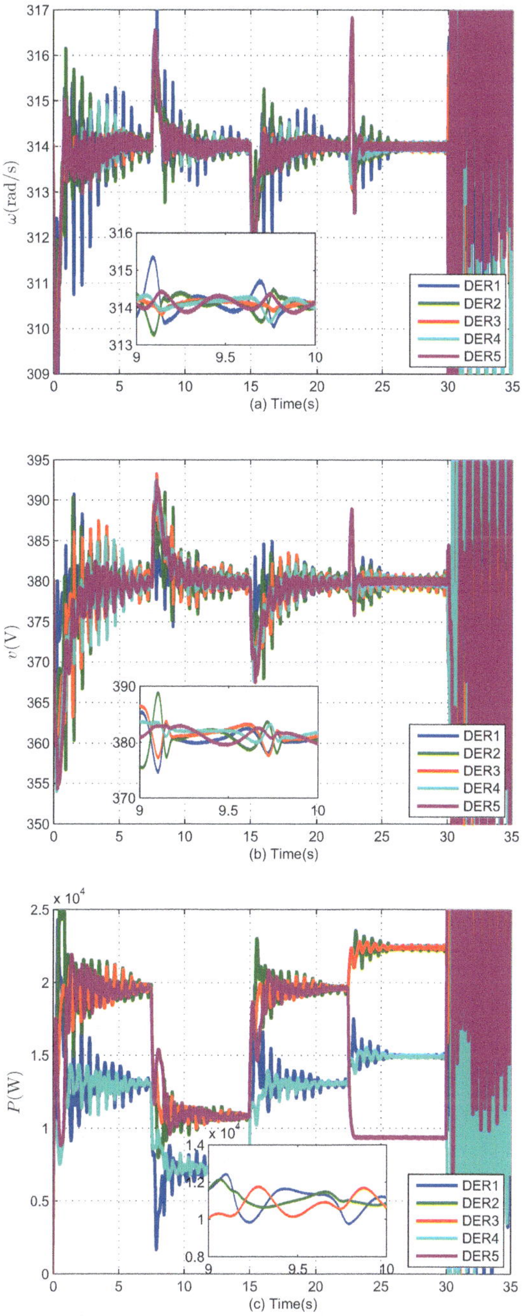

Fig. 6.5 Comparison between the proposed resilient control and the previously proposed control: (**a**) output voltage; (**b**) frequency; and (**c**) optimal active power, subject to additive noise and time delay disturbances (©2024 IEEE. Reprinted, with permission, from IEEE Transactions on Systems, Man, and Cybernetics: Systems [164])

strates consistent stability and precision, highlighting its advantages for microgrid due to communication constraints are inevitable in practice.

6.4 Summary

This chapter has introduced a multi-agent-based resilient control method for AC microgrids. The proposed control strategy ensures synchronous restoration of voltage and frequency for all DER agents and optimal sharing of active power in the presence of additive noises and communication time delays in the AC microgrid network. Under the application of distributed control strategy, there is no need for the intervention of the central processing unit, reducing computational complexity and the risk of single point of failure. With the help of algebraic graph theory and matrix theorem, the leader-follower frequency and voltage synchronous output problem is transformed into a stochastic stability problem in the form of stochastic differential delay. Through rigorous theoretical verification, the stochastic synchronization conditions of the AC microgrid have been derived. Finally, the effectiveness of the proposed scheme has been confirmed through two simulation cases in a simulation verification environment.

Chapter 7
Summary and Outlook

This book has investigated distributed cooperative intelligence in microgrid control system, precisely from the perspective of communication constraints. The utilization of advanced ICTs transforms DERs into intelligent "energy agents" forming a multi-agent system. While this cyber-enhanced, distributed control paradigm offers superior flexibility, scalability, and reliability over conventional centralized structures, its performance is inherently tied to the quality of the sparse communication network. The communication constraints have been central challenges that must be explicitly addressed to realize the full potential of a cooperative microgrid.

Throughout this book, we have systematically investigated the impact of key communication limitations including: finite-time convergence requirements, dynamical network topologies, non-uniform communication time-varying delays, uncertain communication links, limited bandwidth, and stochastic noise disturbances. In response, a series of advanced distributed intelligent cooperative protocols has been proposed:

Chapter 3 introduced a finite-time control protocol, ensuring that critical objectives like frequency restoration and active/reactive power sharing are achieved within finite time. This approach enhances the system's transient performance and steady-state stability, effectively addressing the need for rapid and accurate coordination.

Chapter 4 developed a robust control protocol to handle the challenges of dynamical switching communication networks with non uniform communication time varying delays. By integrating pinning control and consensus based observers, the proposed control scheme maintains system stability and ensures precise voltage regulation and reactive power sharing, provided the network connectivity and delay upper bounds meet certain sufficient conditions.

Chapter 5 proposed a discrete time protocol based on an iterative learning mechanism, specifically designed for operating environments with uncertain communication links. This strategy significantly reduces communication costs by

J. Lai, X. Lu, *Distributed Cooperative Intelligence in Microgrid Control*,
https://doi.org/10.1007/978-981-95-6884-0_7

requiring only intermittent neighbor communication and demonstrates remarkable robustness against internal uncertainties and external measurement noises.

Chapter 6 presented a resilient consensus protocol that guarantees system performance even in the presence of additive noise and communication delay disturbances. By utilizing stochastic analysis, this approach achieves voltage regulation, frequency restoration and optimal power sharing in an almost sure convergence sense, marking a significant step towards reliable operation of microgrids operating in low bandwidth, noisy communication environments.

Collectively, these chapters provide targeted distributed intelligent cooperative strategies to overcome specific communication constraints, thereby reinforcing the stability, security, and intelligence of cyber-physical microgrids.

Future research will delve into more complex and intertwined communication constraints. This includes investigating the effects of nonlinear and hybrid communication delays, malicious cyber-attacks (e.g., false data injection and denial-of-service attacks), and the co-design of communication and control protocols under severely limited bandwidth. Furthermore, the integration of advanced information technologies, such as 5G/6G networks, edge computing, and artificial intelligence will become the focus of our next research. Exploring data-driven and learning-based cooperative control algorithms that can self adapt to evolving network conditions and system parameters will be crucial. Finally, extending these distributed cooperative intelligence frameworks to larger scale, interconnected microgrid clusters will pave the pathway for the next generation of truly intelligent, self-healing, and collaborative energy systems.

Bibliography

1. A. Bidram, A. Davoudi, Hierarchical structure of microgrids control system. IEEE Trans. Smart Grid **3**(4), 1963–1976 (2012)
2. N. Hatziargyriou, H. Asano, R. Iravani, C. Marnay, Microgrids. IEEE Power Energy Magaz. **5**(4), 78–94 (2007)
3. J. Lopes, C. Moreira, A. Madureira, Defining control strategies for microgrids islanded operation. IEEE Trans. Power Syst. **21**(2), 916–924 (2006)
4. J.M. Guerrero, J.C. Vasquez, J. Matas, L.G. de Vicuna, M. Castilla, Hierarchical control of droop-controlled AC and DC microgrids-a general approach toward standardization. IEEE Trans. Ind. Electron. **58**(1), 158–172 (2011)
5. F. Nejabatkhah, Y. W. Li, Overview of power management strategies of hybrid AC/DC microgrid. IEEE Trans. Power Electron. **30**(12), 7072–7089 (2015)
6. S. Liu, X. Li, M. Xia, Q. Qin, X. Liu, Takagi-Sugeno multimodeling-based large signal stability analysis of DC microgrid clusters. IEEE Trans. Power Electron. **36**(11), 12 670–12 684 (2021)
7. B. Ning, Q.-L. Han, L. Ding, A brief overview of recent advances in distributed accelerated secondary control for islanded AC microgrids. IEEE Trans. Industr. Inform. **21**(9), 6605–6614 (2025)
8. T. Morstyn, B. Hredzak, V.G. Agelidis, Control strategies for microgrids with distributed energy storage systems: an overview. IEEE Trans. Smart Grid **9**(4), 3652–3666 (2018)
9. J. Rocabert, A. Luna, F. Blaabjerg, P. Rodrguez, Control of power converters in AC microgrids. IEEE Trans. Power Electron. **27**(11), 4734–4749 (2012)
10. D.E. Olivares, A. Mehrizi-Sani, A.H. Etemadi, C.A. Canizares, R. Iravani, M. Kazerani, A.H. Hajimiragha, O. Gomis-Bellmunt, M. Saeedifard, R. Palma-Behnke, G.A. Jimnez-Estvez, N.D. Hatziargyriou, Trends in microgrid control. IEEE Trans. Smart Grid **5**(4), 1905–1919 (2014)
11. Q.-C. Zhong, G. Weiss, Synchronverters: inverters that mimic synchronous generators. IEEE Trans. Ind. Electron. **58**(4), 1259–1267 (2011)
12. P. Palensky, D. Dietrich, Demand side management: demand response, intelligent energy systems, and smart loads. IEEE Trans. Ind. Electron. **7**(3), 381–388 (2011)
13. J. Sun, Impedance-based stability criterion for grid-connected inverters. IEEE Trans. Power Electron. **26**(11), 3075–3078 (2011)
14. X. Yu, Y. Xue, Smart grids: a cybercphysical systems perspective. Proc. IEEE **104**(5), 1058–1070 (2016)

J. Lai, X. Lu, *Distributed Cooperative Intelligence in Microgrid Control*,
https://doi.org/10.1007/978-981-95-6884-0

15. A.N. Sheta, G.M. Abdulsalam, B.E. Sedhom, A.A. Eladl, Comparative framework for AC-microgrid protection schemes: challenges, solutions, real applications, and future trends. Prot. Control Mod. Power Syst. **8**(2), 1–40 (2023)
16. B. Adineh, R. Keypour, P. Davari, F. Blaabjerg, Review of harmonic mitigation methods in microgrid: from a hierarchical control perspective. IEEE J. Emerging Sel. Top. Power Electron. **9**(3), 3044–3060 (2021)
17. N.S. Coleman, K.L. Ogawa, J. Hill, K.N. Miu, Reconfigurable distribution automation and control laboratory: solar microgrid experiments. IEEE Trans. Power Syst. **33**(6), 6379–6386 (2018)
18. A. Joshi, S. Capezza, A. Alhaji, M.-Y. Chow, Survey on AI and machine learning techniques for microgrid energy management systems. IEEE/CAA J. Autom. Sin. **10**(7), 1513–1529 (2023)
19. S. Maharjan, C. Bai, H. Wang, Y. Yao, F. Ding, Z. Wang, Distribution system blackstart and restoration using ders and dynamically formed microgrids. IEEE Trans. Smart Grid **16**(3), 2100–2114 (2025)
20. Y. Khayat, Q. Shafiee, R. Heydari, M. Naderi, T. Dragicevic, J.W. Simpson-Porco, F. Dorfler, M. Fathi, F. Blaabjerg, J.M. Guerrero, H. Bevrani, On the secondary control architectures of AC microgrids: an overview. IEEE Trans. Power Electron. **35**(6), 6482–6500 (2020)
21. S.A. Darani, C.D. Majhor, W.W. Weaver, R.D. Robinett, O. Abdelkhalik, Optimal positioning of energy assets in autonomous robotic microgrids for power restoration. IEEE Trans. Industr. Inform. **15**(7), 4370–4380 (2019)
22. B. Zhou, J. Zou, C.Y. Chung, H. Wang, N. Liu, N. Voropai, D. Xu, Multi-microgrid energy management systems: architecture, communication, and scheduling strategies. J. Mod. Power Syst. Clean Energy **9**(3), 463–476 (2021)
23. A. Luo, Q. Xu, F. Ma, Y. Chen, Overview of power quality analysis and control technology for the smart grid. J. Mod. Power Syst. Clean Energy **4**(1), 1–9 (2016)
24. J.M. Guerrero, M. Chandorkar, T.-L. Lee, P.C. Loh, Advanced control architectures for intelligent microgrids-part I: decentralized and hierarchical control. IEEE Trans. Ind. Electron. **60**(4), 1254–1262 (2013)
25. F. Katiraei, M. Iravani, Power management strategies for a microgrid with multiple distributed generation units. IEEE Trans. Power Syst. **21**(4), 1821–1831 (2006)
26. C. Wu, Z. Cui, Q. Xia, J. Yue, F. Lyu, An overview of digital twin technology for power electronics: state-of-the-art and future trends. IEEE Trans. Power Electron. **40**(9), 13 337–13 362 (2025)
27. X. Zhuang, H. Li, J. Liu, Y. Wang, J. Luo, Distributed optimal secondary control of microgrids: a consensus-driven hierarchical algorithm with predefined-time convergence. IEEE Trans. Circuits Syst. II Express Briefs **72**(12), 1997–2001 (2025)
28. Y.A.-R.I. Mohamed, E.F. El-Saadany, Adaptive decentralized droop controller to preserve power sharing stability of paralleled inverters in distributed generation microgrids. IEEE Trans. Power Electron. **23**(6), 2806–2816 (2008)
29. Y. W. Li, C.-N. Kao, An accurate power control strategy for power-electronics-interfaced distributed generation units operating in a low-voltage multibus microgrid. IEEE Trans. Power Electron. **24**(12), 2977–2988 (2009)
30. X. Ji, D. Liu, X. Tian, R. Wang, Q. Sun, H. Zhang, Fixed-time quasi-consensus current sharing control for islanded DC microgrids under unknown DoS attacks based on event-triggered control. IEEE Trans. Autom. Sci. Eng. **22**, 23 631–23 643 (2025)
31. X. Liu, P. Wang, P.C. Loh, A hybrid AC/DC microgrid and its coordination control. IEEE Trans. Smart Grid **2**(2), 278–286 (2011)
32. X. She, A.Q. Huang, R. Burgos, Review of solid-state transformer technologies and their application in power distribution systems. IEEE J. Emerging Sel. Top. Power Electron. **1**(3), 186–198 (2013)
33. X. Qiu, B. Fan, W. Meng, Y. Wang, Y. Xu, Z. Dong, Distributed data-driven control for adjustable current sharing and secure voltage restoration in DC microgrids. IEEE Trans. Autom. Sci. Eng. **22**, 23 141–23 151 (2025)

34. L. Ding, Q.-L. Han, X. Ge, X.-M. Zhang, An overview of recent advances in event-triggered consensus of multiagent systems. IEEE Trans. Cybern. **48**(4), 1110–1123 (2018)
35. J. Liu, Y. Miura, T. Ise, Comparison of dynamic characteristics between virtual synchronous generator and droop control in inverter-based distributed generators. IEEE Trans. Power Electron. **31**(5), 3600–3611 (2016)
36. S. Gao, H. Dang, L. Wang, Y. Du, X. Gao, H. Lin, Y. Huangfu, Large-signal stability assessment for islanded MGs with a mix of grid-following and grid-forming inverters: a networked oscillators-based approach. IEEE Trans. Power Electron. **41**(1), 190–196 (2026)
37. W. Wang, R. Chi, Y. Liu, Z. Hou, Model-and-data-driven adaptive frequency control for microgrid systems. IEEE Trans. Autom. Sci. Eng. **22**, 22 655–22 662 (2025)
38. R.H. Lasseter, Smart distribution: coupled microgrids. Proc. IEEE **99**(6), 1074–1082 (2011)
39. X. Lu, J.M. Guerrero, K. Sun, J.C. Vasquez, An improved droop control method for DC microgrids based on low bandwidth communication with DC bus voltage restoration and enhanced current sharing accuracy. IEEE Trans. Power Electron. **29**(4), 1800–1812 (2014)
40. Q. Shafiee, J.M. Guerrero, J.C. Vasquez, Distributed secondary control for islanded microgrids-a novel approach. IEEE Trans. Power Electron. **29**(2), 1018–1031 (2014)
41. H. Guo, X. Dai, S. Bu, Z. Zhang, Distributed secondary control of DC microgrids under unreliable communication networks. IEEE Trans. Autom. Sci. Eng. **22**, 22 900–22 911 (2025)
42. X. Lu, J. Lai, Communication constraints for distributed secondary control of heterogeneous microgrids: a survey. IEEE Trans. Ind. Appl. **57**(6), 5636–5648 (2021)
43. J. de la Cruz, Y. Wu, J.E. Candelo-Becerra, J.C. Vsquez, J.M. Guerrero, Review of networked microgrid protection: architectures, challenges, solutions, and future trends. CSEE J. Power Energy Syst **10**(2), 448–467 (2024)
44. L. Che, M. Shahidehpour, A. Alabdulwahab, Y. Al-Turki, Hierarchical coordination of a community microgrid with AC and DC microgrids. IEEE Trans. Smart Grid **6**(6), 3042–3051 (2015)
45. J. Yu, M. Ni, Y. Jiao, X. Wang, Plug-in and plug-out dispatch optimization in microgrid clusters based on flexible communication. J. Mod. Power Syst. Clean Energy **5**(4), 663–670 (2017)
46. K. Gao, J. Lu, B. Jiang, J. Lou, Enhancing system stability: the impacts of time delays and average impulsive level. IEEE Trans. Autom. Sci. Eng. **22**, 23 476–23 487 (2025)
47. J. Wu, F. Guo, F. Yang, C. Wen, M.-Y. Chow, F. Boem, Resilient net zero: an overview on secure control methods for networked microgrids. IEEE Ind. Electron. Mag. 2–14 (2025). https://doi.org/10.1109/MIE.2025.3624516
48. N.M. Dehkordi, N. Sadati, M. Hamzeh, Distributed robust finite-time secondary voltage and frequency control of islanded microgrids. IEEE Trans. Power Syst. **32**(5), 3648–3659 (2017)
49. J. Liu, J. Li, H. Song, A. Nawaz, Y. Qu, Nonlinear secondary voltage control of islanded microgrid via distributed consistency. IEEE Trans. Energy Convers. **35**(4), 1964–1972 (2020)
50. Z. Li, Z. Cheng, J. Liang, J. Si, L. Dong, S. Li, Distributed event-triggered secondary control for economic dispatch and frequency restoration control of droop-controlled AC microgrids. IEEE Trans. Sustain. Energy **11**(3), 1938–1950 (2020)
51. S.I. Habibi, M.A. Sheikhi, T. Khalili, S.A.G.K. Abadi, A. Bidram, J.M. Guerrero, Multiagent-based nonlinear generalized minimum variance control for islanded AC microgrids. IEEE Trans. Power Syst. **39**(1), 316–328 (2024)
52. Y. Chen, D. Qi, H. Dong, C. Li, Z. Li, J. Zhang, A FDI attack-resilient distributed secondary control strategy for islanded microgrids. IEEE Trans. Smart Grid **12**(3), 1929–1938 (2021)
53. M. Adibi, J. van der Woude, Secondary frequency control of microgrids: an online reinforcement learning approach. IEEE Trans. Autom. Control **67**(9), 4824–4831 (2022)
54. X. Li, Q. Xu, F. Blaabjerg, Adaptive resilient secondary control for islanded AC microgrids with sensor faults. IEEE J. Emerging Sel. Top. Power Electron. **9**(5), 5239–5248 (2021)
55. Z. Deng, Y. Xu, H. Sun, X. Shen, Distributed, bounded and finite-time convergence secondary frequency control in an autonomous microgrid. IEEE Trans. Smart Grid **10**(3), 2776–2788 (2019)

56. Y. Sun, D. Zhao, Y. Li, Y. Zeng, X. Xing, C. Zhang, Data-driven distributed resilient control for DC microgrids with BESSs under DoS attacks and communication delays. IEEE Trans. Transp. Electrif. **11**(6), 13 510–13 521 (2025)
57. L. Meng, F. Tang, M. Savaghebi, J.C. Vasquez, J.M. Guerrero, Tertiary control of voltage unbalance compensation for optimal power quality in islanded microgrids. IEEE Trans. Energy Convers. **29**(4), 802–815 (2014)
58. T. Morstyn, A.V. Savkin, B. Hredzak, H.D. Tuan, Scalable energy management for low voltage microgrids using multi-agent storage system aggregation. IEEE Trans. Power Syst. **33**(2), 1614–1623 (2018)
59. H. Shafei, M. Farhangi, S.K. Sarker, L. Li, R.P. Aguilera, H. Haes Alhelou, A distributed projection operator-based unknown input observer for attack estimation and mitigation on DC microgrids. IEEE Trans. Cybern. **55**(11), 5372–5386 (2025)
60. Z. Liu, L. Xing, J. Fang, Z. Shu, H. Su, Distributed secondary control for DC microgrids with near-infinite constant power load accommodation. IEEE Trans. Smart Grid **16**(6), 4451–4462 (2025)
61. C. Yu, X. Lu, Rocof and nadir of frequency optimization control for cluster-oriented microgrids. IEEE Trans. Smart Grid **16**(6), 4437–4450 (2025)
62. Q. Liu, C. Zhao, X. Luo, P. Cheng, J. Chen, Resilient distributed control for ensuring safety and stability in DC microgrids against false data injection attacks. IEEE Trans. Smart Grid **16**(6), 5486–5499 (2025)
63. Y. Zeng, Q. Zhang, H. Yu, H.H.C. Iu, X. Zhang, Y. Liu, H. Guo, F. Zhang, S. Liu, Multiobjective fusion controller for plug-and-play distributed energy storage systems in shipboard DC microgrids based on dynamic diffusion algorithm. IEEE Trans. Transp. Electrif. **11**(6), 13 351–13 365 (2025)
64. X. Chen, J. Sun, L. Cheng, G. Wen, J. Cao, J. Qiu, Observer-based distributed control and power sharing of multiterminal DC transmission systems with switching topology. IEEE Trans. Syst. Man Cybern.: Syst. **55**(11), 8072–8083 (2025)
65. Y. Wang, T.-L. Nguyen, Y. Xu, Q.-T. Tran, R. Caire, Peer-to-peer control for networked microgrids: multi-layer and multi-agent architecture design. IEEE Trans. Smart Grid **11**(6), 4688–4699 (2020)
66. Y.S.F. Eddy, H.B. Gooi, S.X. Chen, Multi-agent system for distributed management of microgrids. IEEE Trans. Power Syst. **30**(1), 24–34 (2015)
67. S. Alzahrani, K. Sinjari, J. Mitra, Multi-agent and state observer-based technique for microgrid protection. IEEE Trans. Ind. Appl. **60**(2), 2697–2705 (2024)
68. A. Sujil, S.K. Agarwal, R. Kumar, Centralized multi-agent implementation for securing critical loads in PV based microgrid. J. Mod. Power Syst. Clean Energy **2**(1), 77–86 (2014)
69. M.N. Alam, S. Chakrabarti, A. Ghosh, Networked microgrids: state-of-the-art and future perspectives. IEEE Trans. Industr. Inform. **15**(3), 1238–1250 (2019)
70. Z. Li, M. Shahidehpour, F. Aminifar, A. Alabdulwahab, Y. Al-Turki, Networked microgrids for enhancing the power system resilience. Proc. IEEE **105**(7), 1289–1310 (2017)
71. J. Lai, C. Yu, H. Su, Z. Zeng, Distributed secondary frequency cooperation and power allocation in cyber-physical microgrids with multiple operational constraints. IEEE Trans. Cybern. https://doi.org/10.1109/TCYB.2025.3624466 (2025)
72. T.A. Fagundes, R.S. de Castro, M.V.R. Campos, B.M. Zilli, L.J.R. Silva, J.M. Guerrero, R.Q. Machado, A design approach for soc integration with voltage restoration capability for redundancy-based DC microgrids. IEEE Open J. Ind. Electron. Soc. **6**, 1248–1268 (2025)
73. K. Xiahou, K. Zou, Y. Wang, Z. Lin, Y. Liu, Z. Liu, Q. Wu, Distributed observer-based resilient control of cyber-physical DC microgrids against communication-link and local attacks. IEEE Trans. Smart Grid **16**(6), 5459–5471 (2025)
74. C. Yu, H. Zhou, X. Lu, J. Lai, G.-P. Liu, Distributed optimal synchronization rate control for AC microgrids under event-triggered mechanism. IEEE Trans. Power Syst. **36**(3), 1780–1793 (2021)

75. S. Huang, L. Wang, L. Xiong, Y. Zhou, F. Gao, W. Huang, Y. Li, H. Guo, F. Wang, Hierarchical robustness strategy combining model-free prediction and fixed-time control for islanded AC microgrids. IEEE Trans. Smart Grid **16**(6), 4380–4394 (2025)
76. F. Cheng, J. Lai, Z. Zeng, Bilevel distributed secure control of DC microgrid clusters under scale random cyberattacks. IEEE Trans. Smart Grid **16**(6), 4324–4334 (2025)
77. F. Guo, C. Wen, J. Mao, J. Chen, Y.-D. Song, Distributed cooperative secondary control for voltage unbalance compensation in an islanded microgrid. IEEE Trans. Industr. Inform. **11**(5), 1078–1088 (2015)
78. H. Yan, X. Zhou, H. Zhang, F. Yang, Z.-G. Wu, A novel sliding mode estimation for microgrid control with communication time delays. IEEE Trans. Smart Grid **10**(2), 1509–1520 (2019)
79. L. Wei, H. Ye, Q. Mou, D. Pang, M. Liu, X. Shi, Y. Liu, Value set method for robust stability analysis of microgrids with multiple uncertain time delays. IEEE Trans. Control Syst. Technol. **32**(3), 1105–1113 (2024)
80. X. Lu, X. Yu, J. Lai, J.M. Guerrero, H. Zhou, Distributed secondary voltage and frequency control for islanded microgrids with uncertain communication links. IEEE Trans. Ind. Inf. **13**(2), 448–460 (2017)
81. Z. Li, G. Chen, Distributed adaptive control scheme for islanded AC microgrids with tolerance to uncertain communication links. IEEE Syst. J. **16**(2), 2741–2752 (2022)
82. J. Lai, X. Lu, F. Wang, Bilevel information-aware distributed resilient control for heterogeneous microgrid clusters. IEEE Trans. Ind. Appl. **57**(3), 2014–2022 (2021)
83. P. Ge, C. Chen, Y. Wang, Q. Hu, Self-organizing post-contingency distribution network via drone-assisted resilience-oriented control. IEEE Trans. Smart Grid **16**(6), 5528–5540 (2025)
84. P. Prabhakaran, Y. Goyal, V. Agarwal, A novel communication-based average voltage regulation scheme for a droop controlled DC microgrid. IEEE Trans. Smart Grid **10**(2), 1250–1258 (2019)
85. H. Wu, S.-M. Zhu, Z.-H. Zhu, H.-S. Su, G.-P. Wu, W.-B. Zhao, Privacy-preserving adaptive secondary control for DC microgrids with constant power loads. IEEE Trans. Ind. Inf. **21**(10), 7993–8002 (2025)
86. S. Kumar, B. Singh, S. Das, An experimental validation of nonlinear extended state observer controlled small hydro-PV array and battery storage-based microgrid for remote areas. IEEE Trans. Autom. Sci. Eng. **22**, 17 010–17 021 (2025)
87. G.C. Kryonidis, E.O. Kontis, A.I. Chrysochos, K.O. Oureilidis, C.S. Demoulias, G.K. Papagiannis, Power flow of islanded AC microgrids: revisited. IEEE Trans. Smart Grid **9**(4), 3903–3905 (2018)
88. P.C. Loh, D. Li, Y.K. Chai, F. Blaabjerg, Autonomous control of interlinking converter with energy storage in hybrid accdc microgrid. IEEE Trans. Ind. Appl. **49**(3), 1374–1382 (2013)
89. A. Afshari, M. Raeispour, M. Davari, W. Gao, F. Blaabjerg, T. Chai, Finite-rate distributed secondary control over digital communication networks using an event-triggered quantized algorithm for islanded modern microgrids utilizing inverter-based resources. IEEE Trans. Ind. Inf. **21**(2), 1002–1016 (2025)
90. H. Su, J. Lai, Distributed robust iterative learning control for AC microgrids with external disturbances. Sci. China Inf. Sci. **68**(10), 1869–1919 (2025)
91. F. Guo, X. Ye, J. Wu, J. Huang, F. Boem, A decentralized/distributed cluster-based predefined-time control for DC networked microgrids in the presence of constant power loads. Automatica **183**, 112657 (2026)
92. T. Logenthiran, R.T. Naayagi, W.L. Woo, V.-T. Phan, K. Abidi, Intelligent control system for microgrids using multiagent system. IEEE J. Emerging Sel. Top. Power Electron. **3**(4), 1036–1045 (2015)
93. S. Marzal, R. Gonzlez-Medina, R. Salas-Puente, G. Garcer, E. Figueres, An embedded internet of energy communication platform for the future smart microgrids management. IEEE Internet Things J. **6**(4), 7241–7252 (2019)
94. Z. Wang, J. Wang, M. Ma, H. Yang, D. Chen, L. Xiong, P. Li, A distributed event-triggered fixed-time fault-tolerant secondary control framework of islanded AC microgrid against faults and communication constraints. IEEE Trans. Power Syst. **37**(5), 3817–3833 (2022)

95. L. Huang, W. Sun, Q. Li, H. Zhang, W. Li, Optimal communication network design for discrete-time distributed secondary control of islanded microgrid. IEEE Trans. Smart Grid **15**(6), 5346–5357 (2024)
96. J. Schiffer, F. Dorfler, E. Fridman, Robustness of distributed averaging control in power systems: time delays and dynamic communication topology. Automatica **80**, 261–271 (2017)
97. K. Zuo, L. Wu, Eigenvalue-based stability analysis for droop-free controlled islanded microgrid with symmetric/asymmetric communication network. IEEE Trans. Smart Grid **13**(4), 2511–2522 (2022)
98. G. Lou, W. Gu, Y. Xu, W. Jin, X. Du, Stability robustness for secondary voltage control in autonomous microgrids with consideration of communication delays. IEEE Trans. Power Syst. **33**(4), 4164–4178 (2018)
99. M.J. Najafirad, N.M. Dehkordi, M. Hamzeh, H. Nazaripouya, Distributed event-triggered control of DC microgrids with input saturation and time delay constraints. IEEE Syst. J. **17**(3), 4786–4797 (2023)
100. E. Firouzmand, M. Babahaji, H.A. Talebi, A.G. Aghdam, Distributed control of multi-agent systems with uncertain communication links. IEEE J. Radio Freq. Identif. (RFID) **7**, 415–423 (2023)
101. J. Zhong, C. Chen, Z. Bie, M. Shahidehpour, Strategic sdn-based microgrid formation for managing communication failures in distribution system restoration. IEEE Trans. Power Syst. **40**(3), 2506–2518 (2025)
102. J. Lai, X. Lu, A. Monti, R.W. De Doncker, Event-driven distributed active and reactive power dispatch for CCVSI-based distributed generators in AC microgrids. IEEE Trans. Ind. Appl. **56**(3), 3125–3136 (2020)
103. S. Yan, J.H. Park, Bandwidth-aware and anti-outlier fusion-event-driven fault detection of networked photovoltaic microgrids with measurement outliers. IEEE Internet Things J. **12**(23), 50 559–50 568 (2025)
104. M. Alizadeh, H.A. Abyaneh, K. Rouzbehi, An enhanced distributed event-triggered mechanism of cyber-switched communications for control of islanded AC microgrids. IEEE Syst. J. **17**(1), 1501–1511 (2023)
105. N.A. El-Taweel, H.E.Z. Farag, Voltage regulation in islanded microgrids using distributed constraint satisfaction. IEEE Trans. Smart Grid **9**(3), 1613–1625 (2018)
106. J. Lai, X. Lu, X. Yu, A. Monti, H. Zhou, Distributed voltage regulation for cyber-physical microgrids with coupling delays and slow switching topologies. IEEE Trans. Syst. Man Cybern.: Syst. **50**(1), 100–110 (2020)
107. B. Ning, Q.-L. Han, L. Ding, Distributed secondary control of AC microgrids with external disturbances and directed communication topologies: a full-order sliding-mode approach. IEEE/CAA J. Autom. Sin. **8**(3), 554–564 (2021)
108. A. Afshari, M. Karrari, H.R. Baghaee, G.B. Gharehpetian, S. Karrari, Cooperative fault-tolerant control of microgrids under switching communication topology. IEEE Trans. Smart Grid **11**(3), 1866–1879 (2020)
109. X. Lu, J. Lai, Distributed cluster cooperation for multiple DC MGs over two-layer switching topologies. IEEE Trans. Smart Grid **11**(6), 4676–4687 (2020)
110. X. Lu, J. Lai, G.-P. Liu, Mastercslave cooperation for Multi-DC-MGs via variable cyber networks. IEEE Trans. Cybern. **52**(8), 8425–8438 (2022)
111. X. Lu, J. Lai, X. Yu, Y. Wang, J.M. Guerrero, Distributed coordination of islanded microgrid clusters using a two-layer intermittent communication network. IEEE Trans. Ind. Inf. **14**(9), 3956–3969 (2018)
112. Y. Han, K. Zhang, H. Li, E.A.A. Coelho, J.M. Guerrero, MAS-based distributed coordinated control and optimization in microgrid and microgrid clusters: a comprehensive overview. IEEE Trans. Power Electron. **33**(8), 6488–6508 (2018)
113. M. Porfiri, D.J. Stilwell, Consensus seeking over random weighted directed graphs. IEEE Trans. Autom. Control **52**(9), 1767–1773 (2007)

114. S.A. Arefifar, Y.A.-R.I. Mohamed, T.H.M. El-Fouly, Supply-adequacy-based optimal construction of microgrids in smart distribution systems. IEEE Trans. Smart Grid **3**(3), 1491–1502 (2012)
115. H.L. Trentelman, K. Takaba, N. Monshizadeh, Robust synchronization of uncertain linear multi-agent systems. IEEE Trans. Autom. Control **58**(6), 1511–1523 (2013)
116. J. Schiffer, T. Seel, J. Raisch, T. Sezi, Voltage stability and reactive power sharing in inverter-based microgrids with consensus-based distributed voltage control. IEEE Trans. Control Syst. Technol. **24**(1), 96–109 (2016)
117. Z. Hu, S. Liu, W. Luo, L. Wu, Resilient distributed fuzzy load frequency regulation for power systems under cross-layer random denial-of-service attacks. IEEE Trans. Cybern. **52**(4), 2396–2406 (2022)
118. C. Dou, D. Yue, J.M. Guerrero, X. Xie, S. Hu, Multiagent system-based distributed coordinated control for radial DC microgrid considering transmission time delays. IEEE Trans. Smart Grid **8**(5), 2370–2381 (2017)
119. H. Liu, Y. Zhang, J. Su, N. Wang, B. Chen, Distributed optimal control of energy storages in DC microgrids for operation loss minimization with communication delays. IEEE Trans. Smart Grid **15**(6), 5284–5297 (2024)
120. L. Ding, Q.-L. Han, L.Y. Wang, E. Sindi, Distributed cooperative optimal control of DC microgrids with communication delays. IEEE Trans. Ind. Inf. **14**(9), 3924–3935 (2018)
121. S. Yang, Q. Liu, J. Wang, Distributed optimization based on a multiagent system in the presence of communication delays. IEEE Trans. Syst. Man Cybern.: Syst. **47**(5), 717–728 (2017)
122. Z. Artstein, Linear systems with delayed controls: a reduction. IEEE Trans. Autom. Control **27**(4), 869–879 (1982)
123. Y. Yang, C. Hua, J. Li, A novel delay-dependent finite-time control of telerobotics system with asymmetric time-varying delays. IEEE Trans. Control Syst. Technol. **30**(3), 985–996 (2022)
124. G. Chen, C. Fan, J. Sun, J. Xia, Mean square exponential stability analysis for it stochastic systems with aperiodic sampling and multiple time-delays. IEEE Trans. Autom. Control **67**(5), 2473–2480 (2022)
125. X.-C. Shangguan, Y. He, C.-K. Zhang, W. Yao, Y. Zhao, L. Jiang, M. Wu, Resilient load frequency control of power systems to compensate random time delays and time-delay attacks. IEEE Trans. Ind. Electron. **70**(5), 5115–5128 (2023)
126. L. Xing, F. Guo, X. Liu, C. Wen, Y. Mishra, Y.-C. Tian, Voltage restoration and adjustable current sharing for DC microgrid with time delay via distributed secondary control. IEEE Trans. Sustainable Energy **12**(2), 1068–1077 (2021)
127. M. Dong, L. Li, Y. Nie, D. Song, J. Yang, Stability analysis of a novel distributed secondary control considering communication delay in DC microgrids. IEEE Trans. Smart Grid **10**(6), 6690–6700 (2019)
128. R. Zhang, B. Hredzak, Distributed finite-time multiagent control for DC microgrids with time delays. IEEE Trans. Smart Grid **10**(3), 2692–2701 (2019)
129. J. Lai, X. Lu, X. Yu, Stochastic distributed frequency and load sharing control for microgrids with communication delays. IEEE Syst. J. **13**(4), 4269–4280 (2019)
130. J. Zhang, H. Zhang, A high-reliability event-triggered secondary control strategy for microgrid considering time-varying delay. IEEE Syst. J. **17**(1), 617–628 (2023)
131. Z. Zhang, C. Dou, D. Yue, B. Zhang, S. Xu, T. Hayat, A. Alsaedi, An event-triggered secondary control strategy with network delay in islanded microgrids. IEEE Syst. J. **13**(2), 1851–1860 (2019)
132. E.A. Coelho, D. Wu, J.M. Guerrero, J.C. Vasquez, T. Dragicevic, C. Stefanovic, P. Popovski, Small-signal analysis of the microgrid secondary control considering a communication time delay. IEEE Trans. Ind. Electron. **63**(10), 6257–6269 (2016)
133. Y.-P. Tian, S. Chun, G. Chen, S. Zong, Y. Huang, B. Wang, Delay compensation-based time synchronization under random delays: algorithm and experiment. IEEE Trans. Control Syst. Technol. **29**(1), 80–95 (2021)

134. G.P. Liu, Predictive controller design of networked systems with communication delays and data loss. IEEE Trans. Circuits Syst. II Express Briefs **57**(6), 481–485 (2010)
135. G.A. Pagani, M. Aiello, Towards decentralization: a topological investigation of the medium and low voltage grids. IEEE Trans. Smart Grid **2**(3), 538–547 (2011)
136. S. Bolognani, S. Zampieri, A distributed control strategy for reactive power compensation in smart microgrids. IEEE Trans. Autom. Control **58**(11), 2818–2833 (2013)
137. H. Nikkhajoei, R.H. Lasseter, Distributed generation interface to the certs microgrid. IEEE Trans. Power Delivery **24**(3), 1598–1608 (2009)
138. M.A. Mahmud, M.J. Hossain, H.R. Pota, A.M.T. Oo, Robust nonlinear distributed controller design for active and reactive power sharing in islanded microgrids. IEEE Trans. Energy Convers. **29**(4), 893–903 (2014)
139. Y. Seyedi, H. Karimi, J.M. Guerrero, Centralized disturbance detection in smart microgrids with noisy and intermittent synchrophasor data. IEEE Trans. Smart Grid **8**(6), 2775–2783 (2017)
140. S. Abhinav, I.D. Schizas, F.L. Lewis, A. Davoudi, Distributed noise-resilient networked synchrony of active distribution systems. IEEE Trans. Smart Grid **9**(2), 836–846 (2018)
141. Z. Liao, J. Shi, S. Wang, Y. Zhang, R. Mu, Z. Sun, A survey of resilient coordination for cybercphysical systems against malicious attacks. IEEE Internet Things J. **12**(22), 46 500–46 525 (2025)
142. A. Afshari, M. Karrari, H.R. Baghaee, G.B. Gharehpetian, Resilient synchronization of voltage/frequency in AC microgrids under deception attacks. IEEE Syst. J. **15**(2), 2125–2136 (2021)
143. J. Lai, X. Lu, A. Monti, G.-P. Liu, Stochastic distributed pinning control for co-multi-inverter networks with a virtual leader. IEEE Trans. Circuits Syst. II Express Briefs **67**(10), 2094–2098 (2020)
144. N.M. Dehkordi, H.R. Baghaee, N. Sadati, J.M. Guerrero, Distributed noise-resilient secondary voltage and frequency control for islanded microgrids. IEEE Trans. Smart Grid **10**(4), 3780–3790 (2019)
145. J. Lai, X. Lu, Voltage synchronization of discrete-time microgrids over analog fading channels. IFAC-PapersOnLine **53**(2), 12 900–12 905 (2020). 21st IFAC World Congress
146. W. Jin, Y. Zhou, B. Hu, Y. Liu, T. Liu, Privacy-preserving distributed economic dispatch based on state decomposition in smart grids. IEEE Trans. Ind. Inf. **21**(11), 9057–9068 (2025)
147. Q. Duo, H. Junhua, L. Xiaolong, Z. Jiaqiang, Z. Zhihao, Research on consensus of multi-agent systems with and without input saturation constraints. J. Syst. Eng. Electron. **32**(4), 947–955 (2021)
148. X. Li, X. Luo, S. Li, J. Li, X. Guan, Consensus of second-order nonlinear multi-agent systems via sliding mode observer and controller. J. Syst. Eng. Electron. **28**(4), 756–765 (2017)
149. S. Luo, J. Xu, X. Liang, Mean-square consensus of heterogeneous multi-agent systems with time-varying communication delays and intermittent observations. IEEE Trans. Circuits Syst. II Express Briefs **69**(1), 184–188 (2022)
150. S. Su, Z. Lin, Distributed consensus control of multi-agent systems with higher order agent dynamics and dynamically changing directed interaction topologies. IEEE Trans. Autom. Control **61**(2), 515–519 (2016)
151. Z.u.A. Lodhi, K. Zhang, B. Zhou, H. Jiang, Adaptive prescribed-time consensus for a class of nonlinear multi-agent networks by bounded time-varying protocols. IEEE Trans. Circuits Syst. I Regul. Pap. **71**(10), 4742–4752 (2024)
152. G. Zhao, Y. Wang, X. Fu, Hybrid event-triggered consensus tracking of multi-agent systems with discrete control update. IEEE Trans. Circuits Syst. II Express Briefs **69**(4), 2206–2210 (2022)
153. L. Cheng, Y. Wang, W. Ren, Z.-G. Hou, M. Tan, On convergence rate of leader-following consensus of linear multi-agent systems with communication noises. IEEE Trans. Autom. Control **61**(11), 3586–3592 (2016)

154. M. Cao, F. Xiao, L. Wang, Event-based second-order consensus control for multi-agent systems via synchronous periodic event detection. IEEE Trans. Autom. Control **60**(9), 2452–2457 (2015)
155. C. Zhang, X. Dou, J. Zhao, Q. Hu, K. Liu, Y. Lv, Distributed forced-boundary-constrained secondary coordination of islanded microgrids. IEEE Trans. Ind. Appl. **61**(3), 5161–5172 (2025)
156. R. Olfati-Saber, J.A. Fax, R.M. Murray, Consensus and cooperation in networked multi-agent systems. Proc. IEEE **95**(1), 215–233 (2007)
157. H. Li, X. Liao, T. Huang, W. Zhu, Event-triggering sampling based leader-following consensus in second-order multi-agent systems. IEEE Trans. Autom. Control **60**(7), 1998–2003 (2015)
158. Q. Liu, Z. Wang, X. He, D.H. Zhou, Event-based h_∞ consensus control of multi-agent systems with relative output feedback: the finite-horizon case. IEEE Trans. Autom. Control **60**(9), 2553–2558 (2015)
159. Y. Liu, Z. Zuo, J. Song, W. Li, Fixed-time consensus control of general linear multiagent systems. IEEE Trans. Autom. Control **69**(8), 5516–5523 (2024)
160. H. Wang, Q.-L. Han, Distribution of roots of quasi-polynomials of neutral type and its applicationpart II: consensus protocol design of multi-agent systems using delayed state information. IEEE Trans. Autom. Control **69**(6), 4058–4065 (2024)
161. T. Wang, H. Zhang, Y. Zhao, Consensus of multi-agent systems under binary-valued measurements and recursive projection algorithm. IEEE Trans. Autom. Control **65**(6), 2678–2685 (2020)
162. Z. Zhao, Q. Yang, Z. Zhang, Y. Wang, H. Tan, J. Wu, L.L. Lai, Hierarchical distributed model-predictive stabilization control of multi-scale oscillations in windcsolar hybrid multi-microgrids. IEEE Trans. Smart Grid **16**(5), 3476–3488 (2025)
163. X. Lu, F. Austin, S. Chen, Flocking in multi-agent systems with active virtual leader and time-varying delays coupling. Commun. Nonlinear Sci. Numer. Simul. **16**(2), 1014–1026 (2011)
164. J. Lai, X. Lu, Z. Dong, S. Cheng, Resilient distributed multiagent control for AC microgrid networks subject to disturbances. IEEE Trans. Syst. Man Cybern.: Syst. **52**(1), 43–53 (2022)
165. Y. Zhang, Z. Liu, F. Wang, Z. Chen, Distributed power allocation scheme with prescribed performance and intermittent dynamics for besss with discharge rate constraints in microgrids. IEEE Trans. Syst. Man Cybern.: Syst. **55**(3), 1948–1959 (2025)
166. Z.-D. Zhou, G. Guo, R. Zhang, A fixed-time convergent distributed algorithm for time-varying optimal resource allocation problem. IEEE Trans. Signal Inf. Proces. Netw. **11**, 48–58 (2025)
167. F. Cheng, J. Lai, Z. Zeng, Zero-sum game-based distributed secondary control for DC microgrids against stealthy attacks. IEEE Trans. Syst. Man Cybern.: Syst. **55**(2), 864–875 (2025)
168. L. Sun, D. Ding, H. Dong, X. Yi, Distributed economic dispatch of microgrids based on admm algorithms with encryption-decryption rules. IEEE Trans. Autom. Sci. Eng. **22**, 8427–8438 (2025)
169. P.K. Pathak, A.K. Yadav, I. Kamwa, Resilient ratio control assisted virtual inertia for frequency regulation of hybrid power system under DoS attack and communication delay. IEEE Trans. Ind. Appl. **61**(2), 2721–2730 (2025)
170. Q. Wang, S. Huang, L. Xiong, Y. Zhou, T. Niu, F. Gao, M.W. Khan, Z. Wang, C. Ban, R. Song, Distributed secondary control based on Bi-limit homogeneity for AC microgrids subjected to non-uniform delays and actuator saturations. IEEE Trans. Power Syst. **40**(1), 820–833 (2025)
171. Z. Xie, Z. Wu, Output-constrained distributed cooperative control based on dynamic event-triggered average consensus for DC microgrid clusters. CSEE J. Power Energy Syst. **11**(5), 2360–2372 (2025)
172. H. Hua, Y. Zou, X. Chen, K. Yu, L. Gan, S. Lei, J.B. Gertrudes, J. Zheng, D. Sidorov, Dynamic network partitioning for a large-scale microgrid cluster. IEEE Trans. Smart Grid **16**(6), 5615–5627 (2025)

173. L. Gan, H. Chen, L. Hua, Z. Zhang, B. Liu, S. Zhong, Distributed sliding-mode-based resilient secondary restoration control for islanded microgrid under unforeseen influence. IEEE Internet Things J. **12**(20), 42 717–42 727 (2025)
174. A.B. Shyam, S. Anand, S.R. Sahoo, Effect of communication delay on consensus-based secondary controllers in DC microgrid. IEEE Trans. Ind. Electron. **68**(4), 3202–3212 (2021)
175. J. Peng, B. Fan, H. Xu, W. Liu, Discrete-time self-triggered control of DC microgrids with data dropouts and communication delays. IEEE Trans. Smart Grid **11**(6), 4626–4636 (2020)
176. Z. Li, H. Su, Distributed attitude tracking of multiple rigid body systems under an uncertain leader and communication link faults.IEEE Trans. Aerosp. Electron. Syst. **60**(3), 3652–3663 (2024)
177. M. Raeispour, H. Atrianfar, H.R. Baghaee, G.B. Gharehpetian, Robust distributed disturbance-resilient h_{∞}-based control of off-grid microgrids with uncertain communications. IEEE Syst. J. **15**(2), 2895–2905 (2021)
178. S.-W. Lin, C.-T. Ko, B.-Y. Ye, C.-C. Chu, Distributed noise-resilient fixed-time consensus secondary control for hybrid inverter-based AC/DC microgrids. IEEE Trans. Ind. Appl. **59**(6), 7428–7443 (2023)
179. M.R. Khalghani, J. Solanki, S. Khushalani Solanki, A. Sargolzaei, Stochastic secondary frequency control of islanded microgrid under uncertainties. IEEE Syst. J. **15**(1), 1056–1065 (2021)
180. J. Nsengiyaremye, B.C. Pal, M.M. Begovic, Low-cost communication-assisted line protection for multi-inverter based microgrids. IEEE Trans. Power Delivery **36**(6), 3371–3382 (2021)
181. S. Patra, M. Basu, Double-layered droop control-based frequency restoration and seamless reconnection of isolated neighboring microgrids for power sharing. IEEE J. Emerging Sel. Top. Power Electron. **10**(5), 6231–6242 (2022)
182. J. Choi, S.I. Habibi, A. Bidram, Distributed finite-time event-triggered frequency and voltage control of AC microgrids. IEEE Trans. Power Syst. **37**(3), 1979–1994 (2022)
183. F. Ahmadi, Y. Batmani, H. Bevrani, T. Yang, C. Cui, Finite-time synergetic controller design for DC microgrids with constant power loads. IEEE Trans. Smart Grid **14**(5), 3352–3361 (2023)
184. W. Yao, M. Chen, J. Matas, J.M. Guerrero, Z.-M. Qian, Design and analysis of the droop control method for parallel inverters considering the impact of the complex impedance on the power sharing. IEEE Trans. Ind. Electron. **58**(2), 576–588 (2011)
185. J. He, Y.W. Li, An enhanced microgrid load demand sharing strategy. IEEE Trans. Power Electron. **27**(9), 3984–3995 (2012)
186. Y. Zhu, F. Zhuo, F. Wang, B. Liu, R. Gou, Y. Zhao, A virtual impedance optimization method for reactive power sharing in networked microgrid. IEEE Trans. Power Electron. **31**(4), 2890–2904 (2016)
187. H. Mahmood, D. Michaelson, J. Jiang, Accurate reactive power sharing in an islanded microgrid using adaptive virtual impedances. IEEE Trans. Power Electron. **30**(3), 1605–1617 (2015)
188. J. He, Y.W. Li, F. Blaabjerg, An enhanced islanding microgrid reactive power, imbalance power, and harmonic power sharing scheme. IEEE Trans. Power Electron. **30**(6), 3389–3401 (2015)
189. J. Lai, H. Zhou, X. Lu, X. Yu, W. Hu, Droop-based distributed cooperative control for microgrids with time-varying delays. IEEE Trans. Smart Grid **7**(4), 1775–1789 (2016)
190. N. Pogaku, M. Prodanovic, T.C. Green, Modeling, analysis and testing of autonomous operation of an inverter-based microgrid. IEEE Trans. Power Electron. **22**(2), 613–625 (2007)
191. A. Bidram, A. Davoudi, F.L. Lewis, J.M. Guerrero, Distributed cooperative secondary control of microgrids using feedback linearization. IEEE Trans. Power Syst. **28**(3), 3462–3470 (2013)
192. X. Lu, Y. Wang, X. Yu, J. Lai, Finite-time control for robust tracking consensus in mass with an uncertain leader. IEEE Trans. Cybern. **47**(5), 1210–1223 (2017)
193. J.W. Simpson-Porco, Q. Shafiee, F. Derfler, J.C. Vasquez, J.M. Guerrero, F. Bullo, Secondary frequency and voltage control of islanded microgrids via distributed averaging. IEEE Trans. Ind. Electron. **62**(11), 7025–7038 (2015)

194. R. Olfati-Saber, R. Murray, Consensus problems in networks of agents with switching topology and time-delays. IEEE Trans. Autom. Control **49**(9), 1520–1533 (2004)
195. X. Lu, X. Yu, J. Lai, Y. Wang, J.M. Guerrero, A novel distributed secondary coordination control approach for islanded microgrids. IEEE Trans. Smart Grid **9**(4), 2726–2740 (2018)
196. V. Nasirian, S. Moayedi, A. Davoudi, F.L. Lewis, Distributed cooperative control of DC microgrids. IEEE Trans. Power Electron. **30**(4), 2288–2303 (2015)
197. Q.-C. Zhong, Robust droop controller for accurate proportional load sharing among inverters operated in parallel. IEEE Trans. Ind. Electron. **60**(4), 1281–1290 (2013)
198. M. Savaghebi, A. Jalilian, J.C. Vasquez, J.M. Guerrero, Secondary control for voltage quality enhancement in microgrids. IEEE Trans. Smart Grid **3**(4), 1893–1902 (2012)
199. H. Zhang, F.L. Lewis, A. Das, Optimal design for synchronization of cooperative systems: state feedback, observer and output feedback. IEEE Trans. Autom. Control **56**(8), 1948–1952 (2011)
200. P.N. Vovos, A.E. Kiprakis, A.R. Wallace, G.P. Harrison, Centralized and distributed voltage control: impact on distributed generation penetration. IEEE Trans. Power Syst. **22**(1), 476–483 (2007)
201. A.G. Tsikalakis, N.D. Hatziargyriou, Centralized control for optimizing microgrids operation, in *2011 IEEE Power and Energy Society General Meeting* (2011), pp. 1–8
202. A. Vaccaro, G. Velotto, A.F. Zobaa, A decentralized and cooperative architecture for optimal voltage regulation in smart grids. IEEE Trans. Ind. Electron. **58**(10), 4593–4602 (2011)
203. A. Mahajan, Optimal decentralized control of coupled subsystems with control sharing. IEEE Trans. Autom. Control **58**(9), 2377–2382 (2013)
204. Q. Hui, W.M. Haddad, Distributed nonlinear control algorithms for network consensus. Automatica **44**(9), 2375–2381 (2008)
205. C. Zhao, U. Topcu, N. Li, S. Low, Design and stability of load-side primary frequency control in power systems. IEEE Trans. Autom. Control **59**(5), 1177–1189 (2014)
206. H. Bouattour, J.W. Simpson-Porco, F. Drfler, F. Bullo, Further results on distributed secondary control in microgrids, in *52nd IEEE Conference on Decision and Control* (2013), pp. 1514–1519
207. Y. Shan, J. Hu, K.W. Chan, S. Islam, A unified model predictive voltage and current control for microgrids with distributed fuzzy cooperative secondary control. IEEE Trans. Ind. Inf. **17**(12), 8024–8034 (2021)
208. L. Meng, T. Dragicevic, J. Roldn-Prez, J.C. Vasquez, J.M. Guerrero, Modeling and sensitivity study of consensus algorithm-based distributed hierarchical control for DC microgrids. IEEE Trans. Smart Grid **7**(3), 1504–1515 (2016)
209. X. Lu, R. Lu, S. Chen, J. Lu, Finite-time distributed tracking control for multi-agent systems with a virtual leader. IEEE Trans. Circuits Syst. I Regul. Pap. **60**(2), 352–362 (2013)
210. W. Luo, P. Lu, C. Du, H. Liu, Cooperative output tracking control of heterogeneous multi-agent systems with random communication constraints: an observer-based predictive control approach. IEEE Trans. Circuits Syst. II Express Briefs **69**(3), 1139–1143 (2022)
211. X. Wu, Y. Xu, J. He, X. Wang, J. C. Vasquez, J. M. Guerrero, Pinning-based hierarchical and distributed cooperative control for AC microgrid clusters. IEEE Trans. Power Electron. **35**(9), 9865–9885 (2020)
212. W. Yu, G. Chen, J. Lü, On pinning synchronization of complex dynamical networks. Automatica **45**(2), 429–435 (2009)
213. A. Micallef, M. Apap, C. Spiteri-Staines, J.M. Guerrero, J.C. Vasquez, Reactive power sharing and voltage harmonic distortion compensation of droop controlled single phase islanded microgrids. IEEE Trans. Smart Grid **5**(3), 1149–1158 (2014)
214. W. Liu, J. Shen, S. Zhang, N. Li, Z. Zhu, L. Liang, Z. Wen, Distributed secondary control strategy based on q-learning and pinning control for droop-controlled microgrids. J. Mod. Power Syst. Clean Energy **10**(5), 1314–1325 (2022)
215. M. Jamali, H.R. Baghaee, M.S. Sadabadi, G.B. Gharehpetian, A. Anvari-Moghaddam, F. Blaabjerg, Distributed finite-time fault-tolerant control of isolated AC microgrids considering input constraints. IEEE Trans. Smart Grid **13**(6), 4525–4537 (2022)

216. B. Zhao, X. Dong, J. Bornemann, Service restoration for a renewable-powered microgrid in unscheduled island mode. IEEE Trans. Smart Grid **6**(3), 1128–1136 (2015)
217. C.L. Moreira, F.O. Resende, J.A.P. Lopes, Using low voltage microgrids for service restoration. IEEE Trans. Power Syst. **22**(1), 395–403 (2007)
218. H. Xin, Z. Qu, J. Seuss, A. Maknouninejad, A self-organizing strategy for power flow control of photovoltaic generators in a distribution network. IEEE Trans. Power Syst. **26**(3), 1462–1473 (2011)
219. T. Vandoorn, J. De Kooning, B. Meersman, L. Vandevelde, Review of primary control strategies for islanded microgrids with power-electronic interfaces. Renew. Sust. Energ. Rev. **19**, 613–628 (2013)
220. X. Lu, J.M. Guerrero, K. Sun, J.C. Vasquez, R. Teodorescu, L. Huang, Hierarchical control of parallel AC-DC converter interfaces for hybrid microgrids. IEEE Trans. Smart Grid **5**(2), 683–692 (2014)
221. C. Liu, X. Wang, Y. Ren, X. Wang, J. Zhang, A novel distributed secondary control of heterogeneous virtual synchronous generators via event-triggered communication. IEEE Trans. Smart Grid **13**(6), 4174–4189 (2022)
222. P. Buduma, M.K. Das, R.T. Naayagi, S. Mishra, G. Panda, Seamless operation of mastercslave organized AC microgrid with robust control, islanding detection, and grid synchronization. IEEE Trans. Ind. Appl. **58**(5), 6724–6738 (2022)
223. H. Han, X. Hou, J. Yang, J. Wu, M. Su, J.M. Guerrero, Review of power sharing control strategies for islanding operation of AC microgrids. IEEE Trans. Smart Grid **7**(1), 200–215 (2016)
224. Q. Sun, R. Han, H. Zhang, J. Zhou, J.M. Guerrero, A multiagent-based consensus algorithm for distributed coordinated control of distributed generators in the energy internet. IEEE Trans. Smart Grid **6**(6), 3006–3019 (2015)
225. S. Weckx, R. D'Hulst, J. Driesen, Primary and secondary frequency support by a multi-agent demand control system. IEEE Trans. Power Syst. **30**(3), 1394–1404 (2015)
226. W. Liu, W. Gu, W. Sheng, X. Meng, S. Xue, M. Chen, Pinning-based distributed cooperative control for autonomous microgrids under uncertain communication topologies. IEEE Trans. Power Syst. **31**(2), 1320–1329 (2016)
227. A. Merabet, A. Al-Durra, E.F. El-Saadany, Improved feedback control and optimal management for battery storage system in microgrid operating in Bi-directional grid power transfer. IEEE Trans. Sustainable Energy **13**(4), 2106–2118 (2022)
228. H.-S. Ahn, K.L. Moore, Y. Chen, Stability analysis of discrete-time iterative learning control systems with interval uncertainty. Automatica **43**(5), 892–902 (2007)
229. A. Parisio, E. Rikos, L. Glielmo, A model predictive control approach to microgrid operation optimization. IEEE Trans. Control Syst. Technol. **22**(5), 1813–1827 (2014)
230. J. Lai, X. Lu, X. Yu, A. Monti, Cluster-oriented distributed cooperative control for multiple AC microgrids. IEEE Trans. Ind. Inf. **15**(11), 5906–5918 (2019)
231. B. Zhang, C. Dou, D. Yue, J.H. Park, Z. Zhang, Attack-defense evolutionary game strategy for uploading channel in consensus-based secondary control of islanded microgrid considering DoS attack. IEEE Trans. Circuits Syst. I Regul. Pap. **69**(2), 821–834 (2022)
232. K. Zuo, L. Wu, Enhanced power and energy coordination for batteries under the real-time closed-loop, distributed microgrid control. IEEE Trans. Sustainable Energy **13**(4), 2027–2040 (2022)
233. Z. Wang, M. Ma, Q. Zhou, L. Xiong, L. Wang, J. Wang, J. Wang, A privacy-preserving distributed control strategy in islanded AC microgrids. IEEE Trans. Smart Grid **13**(5), 3369–3382 (2022)
234. S. Weng, D. Yue, C. Dou, J. Shi, C. Huang, Distributed event-triggered cooperative control for frequency and voltage stability and power sharing in isolated inverter-based microgrid. IEEE Trans. Cybern. **49**(4), 1427–1439 (2019)
235. Y. Zeng, Q. Zhang, Y. Liu, X. Zhuang, X. Lv, H. Wang, An improved distributed secondary control strategy for battery storage system in DC shipboard microgrid. IEEE Trans. Ind. Appl. **58**(3), 4062–4075 (2022)

236. J. Yang, J. Dai, H.B. Gooi, H.D. Nguyen, P. Wang, Hierarchical blockchain design for distributed control and energy trading within microgrids. IEEE Trans. Smart Grid **13**(4), 3133–3144 (2022)
237. S.M. Mohiuddin, J. Qi, Optimal distributed control of ac microgrids with coordinated voltage regulation and reactive power sharing. IEEE Trans. Smart Grid **13**(3), 1789–1800 (2022)
238. R. Zhang, B. Hredzak, Distributed control system with aperiodic sampled time-delayed data for batteries and renewable energy sources in microgrid. IEEE Trans. Sustainable Energy **11**(2), 1013–1022 (2020)
239. M. Shi, X. Chen, J. Zhou, Y. Chen, J. Wen, H. He, Distributed optimal control of energy storages in a DC microgrid with communication delay. IEEE Trans. Smart Grid **11**(3), 2033–2042 (2020)
240. X. Lu, J. Lai, X. Yu, A novel secondary power management strategy for multiple AC microgrids with cluster-oriented two-layer cooperative framework. IEEE Trans. Ind. Inf. **17**(2), 1483–1495 (2021)
241. G. Chen, Z. Guo, Distributed secondary and optimal active power sharing control for islanded microgrids with communication delays. IEEE Trans. Smart Grid **10**(2), 2002–2014 (2019)
242. S. Liu, X. Wang, P.X. Liu, Impact of communication delays on secondary frequency control in an islanded microgrid. IEEE Trans. Ind. Electron. **62**(4), 2021–2031 (2015)
243. J. Lai, X. Lu, Nonlinear mean-square power sharing control for AC microgrids under distributed event detection. IEEE Trans. Ind. Inf. **17**(1), 219–229 (2021)
244. Y. Xie, Z. Lin, Distributed event-triggered secondary voltage control for microgrids with time delay. IEEE Trans. Syst. Man Cybern.: Syst. **49**(8), 1582–1591 (2019)
245. Y. Du, X. Lu, B. Chen, F. Lin, Resiliency augmented hybrid AC and DC distribution systems with inverter-dominated dynamic microgrids. IEEE Trans. Smart Grid **13**(5), 4088–4101 (2022)
246. G. Liu, Q. Sun, R. Wang, H. Zhang, Event-based fuzzy adaptive consensus FTC for microgrids with nonlinear item via prescribed fixed-time performance. IEEE Trans. Circuits Syst. I Regul. Pap. **69**(7), 2982–2993 (2022)
247. P. Chen, S. Liu, B. Chen, L. Yu, Multi-agent reinforcement learning for decentralized resilient secondary control of energy storage systems against DoS attacks. IEEE Trans. Smart Grid **13**(3), 1739–1750 (2022)
248. L. Yan, M. Sheikholeslami, W. Gong, M. Shahidehpour, Z. Li, Architecture, control, and implementation of networked microgrids for future distribution systems. J. Mod. Power Syst. Clean Energy **10**(2), 286–299 (2022)
249. B. Ning, Q.-L. Han, L. Ding, Distributed finite-time secondary frequency and voltage control for islanded microgrids with communication delays and switching topologies. IEEE Trans. Cybern. **51**(8), 3988–3999 (2021)
250. B. Huang, L. Liu, H. Zhang, Y. Li, Q. Sun, Distributed optimal economic dispatch for microgrids considering communication delays. IEEE Trans. Syst. Man Cybern.: Syst. **49**(8), 1634–1642 (2019)
251. J. Lai, X. Lu, X. Yu, A. Monti, Stochastic distributed secondary control for AC microgrids via event-triggered communication. IEEE Trans. Smart Grid **11**(4), 2746–2759 (2020)
252. J. Hu, G. Feng, Distributed tracking control of leader-follower multi-agent systems under noisy measurement. Automatica **46**(8), 1382–1387 (2010)
253. G. Chen, F.L. Lewis, E.N. Feng, Y. Song, Distributed optimal active power control of multiple generation systems. IEEE Trans. Ind. Electron. **62**(11), 7079–7090 (2015)

Index

J. Lai, X. Lu, *Distributed Cooperative Intelligence in Microgrid Control*,
https://doi.org/10.1007/978-981-95-6884-0

The manufacturer's authorised representative in the EU is Springer Nature Customer Service Centre GmbH, Europaplatz 3, 69115 Heidelberg, Germany. If you have any concerns regarding our products, please contact ProductSafety@springernature.com

Printed and bound by CPI Group (UK) Ltd, Croydon, CR0 4YY
07/10/2026
02229349-0001